VIRUS [illegible] N

Viruses do not behave like other microbes; their life cycles require infecting healthy cells, commandeering their cellular apparatus, replicating, and then killing the host cell. Methods for virus detection and identification have been developed only in the past few decades. These recently developed methods include molecular, physical, and proteomic techniques. All these approaches (electron microscopy, molecular, direct counting, and mass spectrometry proteomics) to detection and identification are reviewed in this succinct volume. It is written in approachable language with enough detail for trained professionals to follow and want to recommend to others.

Key Features

- Covers common detection methods
- Reviews the history of detection from antiquity to the present
- Documents the strengths and weaknesses of various detection methods
- Describes how to detect newly discovered viruses
- Recommends specific applications for clinical, hospital, environmental, and public health uses.

POCKET GUIDES TO
BIOMEDICAL SCIENCES

Series Editor

Lijuan Yuan

A Guide to AIDS, *by Omar Bagasra and Donald Gene Pace*

Tumors and Cancers: Central and Peripheral Nervous Systems, *by Dongyou Liu*

A Guide to Bioethics, *by Emmanuel A. Kornyo*

Tumors and Cancers: Head – Neck – Heart – Lung – Gut, *by Dongyou Liu*

Tumors and Cancers: Skin – Soft Tissue – Bone – Urogenitals, *by Dongyou Liu*

Tumors and Cancers: Endocrine Glands – Blood – Marrow – Lymph, *by Dongyou Liu*

A Guide to Cancer: Origins and Revelations, *by Melford John*

Pocket Guide to Bacterial Infections, *edited by K. Balamurugan*

A Beginner's Guide to Using Open Access Data, *by Saif Aldeen Saleh Airyalat and Shaher Momani*

Pocket Guide to Mycological Diagnosis, *edited by Rossana de Aguiar Cordeiro*

Genome Editing Tools: A Brief Overview, *by Reagan Mudziwapasi and Ringisai Chekera*

Vaccine Efficacy Evaluation: The Gnotobiotic Pig Model, *by Lijuan Yuan*

The Guinea Pig Model: An Alternative Method for Vaccine Potency Testing, *by Viviana Parreño*

Virus Detection, *by Charles H. Wick*

For more information about this series, please visit:

https://www.crcpress.com/Pocket-Guides-to-Biomedical-Sciences/book-series/CRCPOCGUITOB

VIRUS DETECTION

Charles H. Wick

CRC Press is an imprint of the
Taylor & Francis Group, an **informa** business

First edition published 2023
by CRC Press
6000 Broken Sound Parkway NW, Suite 300, Boca Raton, FL 33487-2742
and by CRC Press
4 Park Square, Milton Park, Abingdon, Oxon, OX14 4RN

CRC Press is an imprint of Taylor & Francis Group, LLC

ISBN: 978-0-367-61801-8 (hbk)
ISBN: 978-0-367-61798-1 (pbk)
ISBN: 978-1-003-10662-3 (ebk)

DOI: 10.1201/9781003106623

Typeset in Frutiger
by Deanta Global Publishing Services, Chennai, India

This work is dedicated to:

Virginia Mason Morris Wick

Contents

Series Editor Introduction xi
Pocket Guides to Biomedical Sciences xiii
Preface xv
Acknowledgments xvii
Author Biography xix
List of Abbreviations and Glossary xxi

1 Civilization and Disease **1**

2 Microbes, Fungi, Bacteria, and Viruses **21**
2.1 Fungi 23
2.1.1 What is a fungus? 23
2.1.2 How are fungi detected/classified? 23
2.2 Bacteria 24
2.2.1 What are bacteria? 24
2.2.2 How are bacteria detected and classified? 24
2.3 Viruses 25
2.3.1 What are viruses? 25
2.3.2 How are viruses detected and classified? 25

3 Indirect Methods of Detecting Viruses **33**
3.1 Introduction 33

4 Electron Microscopy **35**
4.1 Introduction 35
4.2 Transmission electron microscopy 36
4.2.1 How does TEM work? 36
4.2.2 How do you use electron microscopy? 37
4.2.3 How do you identify a new virus? 38
4.3 Scanning electron microscopy (SEM) 38
4.3.1 How does scanning electron microscopy work? 38
4.3.2 How do you use scanning electron microscopy? 38
4.3.3 How do you identify a new virus? 43
4.4 Illustrations of viruses based on TEM and SEM visualizations 43

5 Molecular Methods for Detecting Viruses **45**
5.1 Introduction 45
5.2 Polymerase chain reaction (PCR) 46
5.2.1 How do you add new viruses? 51
5.2.1.1 Examples of PCR instruments 51

5.3 Discussion of PCR methods 53
5.4 Antibody methods 54
5.4.1 Detecting viruses using antibodies and how do they work? 54
5.5 How do you add new viruses to the antibody method of detection? 55

6 Direct Virus Counting Methods, Such as IVDS 57
6.1 Introduction 57
6.2 How does direct counting work? 58
6.3 How do you identify a new virus with direct counting? 58
6.4 Why IVDS was invented 58
6.5 Flow chart showing how to use IVDS for virus detection 60
6.6 The recommended uses of IVDS 62
6.7 Improving the sensitivity of IVDS (concentration and accumulation) 66
6.8 An example – following COVID-19 through 5 days and then a 3-month follow-up 66
6.9 PCR and IVDS compared 76
6.10 Summary of the fielded IVDS 77

7 Mass Spectrometry Proteomic (MSP) Method 79
7.1 Introduction 79
7.2 Ion mobility and various types of mass analyzers 81
7.2.1 Using the electrospray ionization (ESI) method in detecting viruses 82
7.3 How do MSP methods work for biological detection? 83
7.4 Detection and identification of viruses using MSP 84
7.5 Examples of viruses detected by MSP methods using ABOid 87
7.5.1 African swine fever virus (ASFV) – Variola porcina 87
7.5.2 *Alcelaphine herpesvirus 1* (AlHV-1) 88
7.5.3 Camelpox virus (CMLV) 89
7.5.4 *Cercopithecine herpesvirus 5* (CeHV-5) 89
7.5.5 Goatpox virus Pellor (GTPV) 89
7.5.6 Lumpy skin disease virus (LSDV) 89
7.5.7 Monkeypox virus Zaire-96-I-16 (MPV) 90
7.5.8 Sheeppox virus 90
7.5.9 Vaccinia virus (VACV) 90
7.5.10 Variola virus (VARV) 90
7.5.11 Discussion of viruses detected by MSP and ABOid 91
7.6 Adding new viruses using MSP 92
7.7 Coronavirus detection including SARS 95
7.7.1 Coronaviruses 95
7.7.2 National average for coronavirus 99
7.7.2.1 Verifying COVID-19 detection 99
7.7.2.2 COVID-19 detection discussion 99
7.8 Summary of MSP-ABOid detection of viruses 99

8 Discussion **101**
8.1 Introduction 101
8.1.1 Proven science 103
8.1.2 Accurate detection 104
8.1.3 Affordable 104
8.1.4 Screens for unknown viruses 106
8.1.5 Ability to detect multiple viruses 106
8.1.6 Quick results (5–10 min) 106
8.2 What are the challenges when detecting viruses? 108
8.2.1 Interference 108
8.2.2 Sensitivity or trust in a particular technology 108
8.2.3 Other things 109
8.3 Clinical 109
8.3.1 Clinic 109
8.3.2 Centralized testing center – separate from a hospital 110
8.3.3 Hospital 110
8.4 Environmental 110
8.4.1 Agriculture – insects (bees), plants, and animals 110
8.4.2 Water 112
8.4.3 Research 112
8.5 Use during a pandemic 113
8.6 Other uses 115
8.6.1 Public use 115
8.6.2 Fixed sites 115
8.6.3 Protecting small high value groups 116
8.6.4 Protecting small groups 116
8.6.5 Protecting large groups 117

References 119
Index 133

Series Editor Introduction

Lijuan Yuan is a Professor of Virology and Immunology in the Department of Biomedical Sciences and Pathobiology, Virginia-Maryland College of Veterinary Medicine, Virginia Tech. Dr. Yuan studies the interactions between enteric viruses and the host immune system. Her lab's research interests are focused on the pathogenesis and immune responses induced by enteric viruses, especially noroviruses and rotaviruses, and on the development of safer and more effective vaccines as well as passive immune prophylaxis and therapeutics against viral gastroenteritis. These studies utilize wild-type, gene knock-out, and human gut microbiota transplanted gnotobiotic pig models of human rotavirus and norovirus infection and diseases. Dr. Yuan's research achievements over the past 29 years have established her as one of a few experts in the world leveraging the gnotobiotic pig model to study human enteric viruses and vaccines. She is the author of the book titled *Vaccine Efficacy Evaluation: The Gnotobiotic Pig Model* in this Pocket Guides to Biomedical Sciences Series. Currently, the Yuan lab is evaluating the immunogenicity and protective efficacy of several candidate novel rotavirus and norovirus vaccines and engineered probiotic yeast *Saccharomyces boulardii* secreting multi-specific single-domain antibodies as novel prophylaxis against both noroviruses and *Clostridioides difficile* infection. See: https://vetmed.vt.edu/people/faculty/yuan-lijuan.html

Pocket Guides to Biomedical Sciences

Book series preface

The *Pocket Guides to Biomedical Sciences* series is designed to provide concise, state-of-the-art, and authoritative coverage on topics that are of interest to undergraduate and graduate students of biomedical majors, health professionals with limited time to conduct their own literature searches, and the general public who are seeking reliable, trustworthy information in biomedical fields. Since its inauguration in 2017, the series has published 12 books (https://www.routledge.com/Pocket-Guides-to-Biomedical-Sciences/book-series/CRCPOCGUITOB) that cover different areas of biomedical sciences. The recent two titles form unique sister pair volumes *Vaccine Efficacy Evaluation: The Gnotobiotic Pig Model* and *The Guinea Pig Model: An Alternative Method for Vaccine Potency Testing*. In these two books, the authors reviewed their decades-long research efforts in the development of two unconventional animal models for vaccine development, evaluation, and quality control. Testing the immunogenicity, protective efficacy, and safety in animal models is one of the most important steps in vaccine development after the construction and formulation of the protective antigens and before human clinical trials. Pig (*Sus scrofa*) has high similarities with humans in gastrointestinal anatomy, physiology, nutritional/dietary requirements, and mucosal immunity. For pre-clinical testing of human rotavirus and norovirus vaccines, an animal model that can exhibit the same or similar clinical signs of disease as humans is critical for assessing protection against both infection and disease upon challenge. Gnotobiotic pig models fulfill this need. Mouse models are more readily available than pig models and are useful for testing vaccine immunogenicity; however, mice cannot be infected by human rotavirus or human norovirus and are not useful for evaluation of vaccine-induced adaptive immunity associated with protection against human rotavirus or norovirus disease. Through the author's studies detailed in her book, the gnotobiotic pig model has been firmly established as the most reliable animal model for the preclinical evaluation of human rotavirus or human norovirus vaccines. Equally important, but for animal vaccines, the guinea pig (*Cavia porcellus*) model represents an alternative method for testing the potency of viral vaccines applied in cattle.

Due to the robustness of the statistical validation, the guinea pig model has been adopted by the National Service of Animal Health of Argentina (sanitary resolution 598.12) as the official potency testing model. The recommendation and guidelines to apply the guinea pig model in the quality control of Infectious Bovine Rhinotracheitis (IBR, caused by bovine herpesvirus type 1), rotavirus, and parainfluenza vaccines were agreed by all the member countries of the American Committee for Veterinary Medicines (CAMEVET), focal point of the World Organization for Animal Health (OIE) in the Americas. These two books demonstrate the indispensable role of animal models in biomedical research and in developing and producing efficacious vaccines that are critical for improving human and animal health. The current title *Virus Detection* describing the four principal methods to detect viruses is a very timely issue. Testing of virus shedding is one of the key strategies to control transmission of viral diseases. The significance of this pocket guide book is highlighted by the recent emergence and reemergence of high impact human viral diseases worldwide.

Preface

The four main methods for detecting viruses are (1) electron microscopy, (2) biomolecular, (3) physical, and (4) mass spectrometry proteomics. The first is detection using an electron microscope which was invented in the 1930s and has become a standard instrument for visualizing the submicron world. It can see all the known viruses and easily functions in the range from 10 to 500 nm. From the pictures we get graphics that illustrate the features of different viruses and we can determine the size. In this manner we can see that viruses are of different sizes starting with polio which is around 20 nm to dengue fever virus at 50 nm to influenza at 90 nm for influenza A and 102 nm for influenza B. Smallpox is seen at around 250 nm. The second are the many biomolecular methods which are based on the genome of the virus. After the discovery of DNA there was an explosion of discovery and organisms of all sorts were classified by their genomic relationship and a phylogenic tree resulted showing these relationships. The viruses were included and more than 44,000 viruses have been sequenced and classified. Using this information related groups or individual viruses could be detected by their genetic information. PCR was invented in the 1990s and has become an important method for detecting viruses based on a portion of their genetic sequence. Additionally, antibodies which are the result of a viral infection can be used to detect when a virus is present. The third method is direct counting of whole viruses by the use of a differential mobility analyzer (DMA). IVDS, invented in the late 1990s, is the first instrument able to detect multiple viruses according to size in a few minutes. The advantages of this method are that there are no reagents, minimum sample prep, and the ability to detect viruses which are not sequenced. The last method, invented in the 2000s, is the mass spectrometer proteomics (MSP) method. This is a method which uses software to detect the unique peptides associated with the virus to determine a detection and classification.

Each of these main detection methods has spun off several variations in their use, but other applications rely on their basic science. Volumes have been written on how these methods are used and their particular manner, such as electron micrography. In this book, each of these methods is presented and discussed as they apply to the detection of viruses. How these methods can be used to complement each other is also discussed. Each method continues to improve as technology improves, but it is important to realize the limitations inherent in each method so as not to be caught unaware.

The world has made great strides in technology. New weapons have likewise improved, we need to be aware of how to counter them, particularly new viruses. It is now possible to manufacture a virus. A virus can be made, adjusted to have a different function or capability, and used for good as in the delivery of medication or for insect control. Viruses can easily be made for malicious purposes. New diseases can be made which are not sequenced, can mimic other viruses in genetic information (appear as benign), and be invisible to most detection methods.

Reading this book should raise your understanding of the methods for detecting viruses. It should give the reader enough information to ask questions about the methods that people are using and more importantly to know the limitations of each method.

Acknowledgments

Thank you to all the people who have asked questions on how to detect viruses. Special thank you to David A. Wick for his many conversations and dedication to detecting viruses and to Jinnie A. Wick, graphic designer, who finished the many figures and proof read the text. Thank you to Sam and Van at All digital for their patience and the many printings.

Author Biography

Dr. Charles H. Wick is a retired senior scientist from the US Army Edgewood Chemical Biological Center (ECBC) where he served both as a manager and research physical scientist and has made significant contributions to forensic science. Although his 40-year professional career has spanned both the public sector and the military, his better-known work in the area of forensic science has occurred in concert with the Department of Defense.

Dr. Wick earned four degrees from the University of Washington and worked in the private sector (civilian occupations) for 12 years, leading to a patent, numerous publications, and international recognition among his colleagues.

In 1983, Dr. Wick joined the Vulnerability/Lethality Division of the United States Army Ballistic Research Laboratory, where he quickly achieved recognition as a manager and principal investigator. It was at this point that he made one of his first major contributions to forensic science and to the field of antiterrorism; his team was the first to utilize current technology to model sub-lethal chemical, biological, and nuclear agents. This achievement was beneficial to all areas of the Department of Defense, as well as to the North Atlantic Treaty Organization (NATO), and gained Wick international acclaim as an authority on individual performance for operations conducted on a nuclear, biological, and chemical (NBC) battlefield.

During his career in the United States Army, Wick rose to the rank of Lieutenant Colonel in the Chemical Corps. He served as a Unit Commander for several rotations, a staff officer for six years (he was a Division Chemical Staff Officer for two rotations), Deputy Program Director Biological Defense Systems, and retired from the position of Commander of the 485th Chemical Battalion in April of 1999.

Dr. Wick continued to work for the DOD as a civilian at the ECBC. Two notable achievements, and one which earned him the Department of the Army Research and Development Award for Technical Excellence and a Federal Laboratory Consortium Technology Transfer Award in 2002, include his invention of the Integrated Virus Detection System (IVDS), a fast-acting, highly portable, user-friendly, extremely accurate, and efficient system for detecting the presence of viruses for the purpose of detection, screening, and characterization. The IVDS can detect the full spectrum of known, unknown, and mutated viruses. This system is compact, portable, and does not rely upon elaborate chemistry. The second, and equally award winning, was his creation and leadership in the development of software designed for detecting and identifying microbes using mass spectrometry proteomics. Each of these projects represents determined ten-year efforts and is novel in its approach to the detection and classification of microbes from complex matrices and the subjects included in three of his books published by CRC Press (*Identifying Microbes by Mass Spectrometry Proteomics*, 2013, *Integrated Virus Detection*, 2014, *Microbial Diversity in Honeybees*, with D.A. Wick, 2021).

Throughout his career, Dr. Wick has made lasting and important contributions to forensic science and to the field of antiterrorism. Dr. Wick holds several US patents in the area of microbe detection and classification. He has written more than 45 civilian and military publications and has received myriad awards and citations, including the Department of the Army Meritorious Civilian Service Medal, the Department of the Army Superior Civilian Service Award, two United States Army Achievement Medals for Civilian Service, the Commander's Award for Civilian Service, the Technical Cooperation Achievement Award, and 25 other decorations and awards for military and community service.

List of Abbreviations and Glossary

Bacteriophage	Virus that attacks bacteria
Cross flow	Sweeping action created by fluid flow across a membrane
CPC	Condensate particle counter
Dalton (D)	Molecular weight of macromolecules
Diafiltration	Dialysis type of filtration
Dialysis	Separation of salts and micro solutes from macromolecular solutions by a concentration gradient
DMA	Differential mobility analyzer
Electrospray (ES)	Process of atomizing a liquid by injecting across an electrical potential
GEMMA	Gas-phase electrophoretic mobility molecular analyzer
IVDS	Integrated Virus Detection System
MW	Molecular weight
Phage	See Bacteriophage
Plaque forming unit (pfu)	Quantity of a virus required to form a viable colony
MS2	Type of bacteriophage
MWCO	Molecular weight cut-off
SMPS	Scanning mobility particle sizer
T7	Type of bacteriophage
Tangential flow	See Cross flow
Ultrafiltration (UF)	Separation of salts and micro solutes from macromolecular solutions by hydrostatic pressure

1 Civilization and Disease

Disease has been with civilization for a long time. Many ancient cultures had outbreaks of disease affecting large numbers of people. We do not know for certain that these outbreaks were caused by any particular organism, but we can look at what was written and make assumptions based on the symptoms and related information. What is not talked about is the fact that the people lived in a biological soup, a cloud of microbes, of which they were unaware. Disease may have been caused by any number of different microbes. The ancient sewage systems, as they were called, did nothing to improve the exposure to microbes. People often had infections that associated with wounds, scratches, teeth, and other common aliments known to modern science, but unknown to the ancient world.

The beginning of the viruses and civilization story starts many thousands of years ago when mankind first organized into groups, towns, cities, and civilizations. It is likely that the association of viruses with people started earlier, but that is largely lost in time and is part of the fossil record. Needless to say, when civilization began so did the plagues and other less common ravages of people by microbes.

The world population was generally less than 50 million until around the first century of the common era when for various reasons it started to increase until the present, when it totals more than 7.5 billion people (Figure 1.1). Associated with this was life expectancy. Generally, following the same curve as the global population, people lived an average of 20–25 years for tens of thousands of years, until the modern era when it increased to the present 73 years (Figure 1.2). Now, considering all the variables and exceptions, the trend was a young and less than 50 million population for many thousands of years and this changed only in the last 2000 years. In the past people did not expect to grow old; in the present most people expect old age and often see life as only beginning in the late 40s and 50s. This perception has changed our relationship to disease.

Considering the average age of people, the total population, and all the associated problems of civilization, an understanding of the attitudes towards disease crystalizes. Average people were occupied with civilization and not really concerned about something they could not see, smell

DOI: 10.1201/9781003106623-1

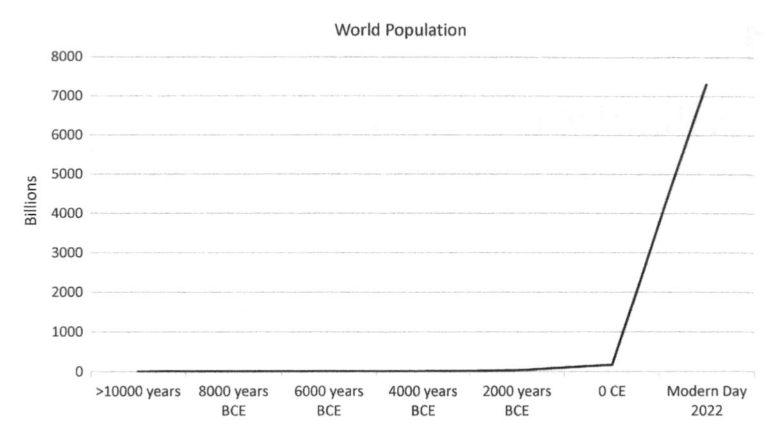

Figure 1.1 Estimated world population from 10,000 years ago to modern times.

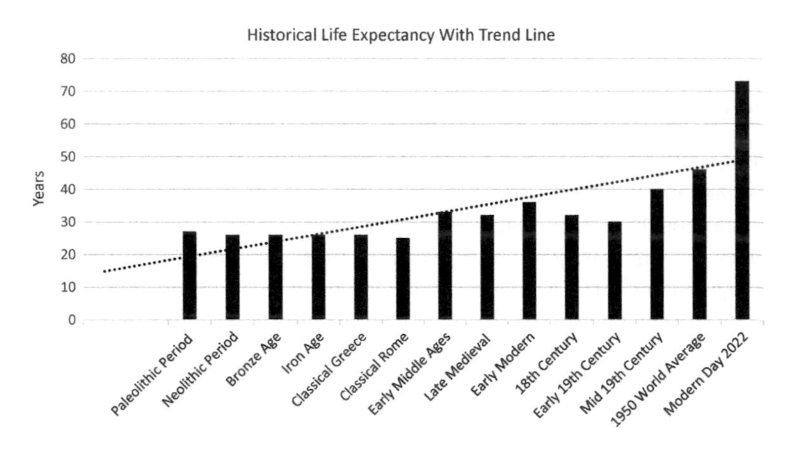

Figure 1.2 Life expectancy from Paleolithic times to modern times.

or understand – other than perhaps in metaphysical terms. The viruses had unlimited access to a susceptible population and simply followed their biology to ravage the population from time to time. Let us look at the recorded or suspected occurrences over the last thousands of years.

It is important to remember that ancient people had little knowledge of the germ theory of disease proposed by Girolamo Fracastoro in 1546, and expanded upon by Marcus von Plenciz in 1762. Nevertheless, these views were not considered creditable by most people who held onto the thoughts of the ancient Greeks and the teaching of Galen who proposed the idea of *miasma* or "bad air". This remained the dominant belief for centuries, particularly among medical people and scientists, until the modern age.

Before 1546–1762, nothing really explained what or why disease occurred, disease was just a part of life and common in the ancient world. People lived together in tight groups to protect themselves from predators and each other. Hygiene was poor and there was no refrigeration to keep food fresh, and living conditions were bleak with the average life expectancy of only about 20–23 years, but people managed to survive.

Let us look at the recorded or suspected occurrences of disease over the last thousands of years. During the Paleolithic age, about three million years ago, people lived as small isolated hunter gatherer populations grouped together in small bands. They survived by gathering plants, fishing, hunting, and scavenging for food. At this time people started to use crude knapped stone tools as well as those made of wood and bone. Neolithic people of 12,000 years ago were the first to develop the use of farming; civilizations began to emerge in the Bronze Age and Iron Age. Based what these early people ate and drank, if they retained their teeth and did not perish as an infant or during childbirth they lived longer. Infections and disease were part of their everyday life; people lived and died.

As people created more densely populated agricultural communities, viruses and all sorts of microbes were allowed to spread rapidly and become endemic within the populations. Livestock and plant viruses were indicated and increased as humans developed and became dependent on agriculture and farming. Diseases such as influenza, potyviruses, rinderpest, and poxviruses of cattle, pigs, and sheep were common.

Microbes are not thinking predators in the ordinary sense; they are opportunists and survive by having an ability to multiply in a large range of environments; people and their livestock provide a good environment.

Around 9500 BCE humans became farmers, and they tended to create monoculture agriculture, growing one type of plant, or several plants of the same species. This development led to the spread of several plant viruses which emerged in fruits and vegetables including the sobemovirus or the Southern bean mosaic virus of potatoes. Measles and smallpox viruses are among the oldest known viruses that infected humans in prehistory. These first viruses are thought to have infected only animals but due to domestication and their close proximity to humans these viruses started to infect humans, first appearing in Europe and African communities thousands of years ago.

The nature of viruses and what they looked like was unknown until the invention of the electron microscope in the 1930s and the discovery of DNA by Watson and Creek in the 1950s. This provided the tools to look at what was previously invisible, and the study and understanding of viruses began. By this discovery many of the diseases reported were found to be caused by viruses in retrospect; for example, polio caused by the poliomyelitis virus had devasted populations in the time of ancient Egypt.

With these modern inventions and later biomolecular methods such as polymerase chain reaction (PCR) we now know viruses played important roles in ecosystems and are essential to life throughout history by creating a change in evolution by transferring genes across species. Over the past 50–100 thousand years as humans migrated throughout the world and populated vast new areas of all kinds of environments, whether land in the tropics, coastal areas, deserts, or hot or cold areas, they carried with them and were exposed to new diseases, including viruses. Most viral epidemics did not exist in early human areas because they lived in small, isolated communities. Smallpox, however, first appeared in larger agricultural communities in India 11,000 years ago. It is thought that these poxviruses first descended from rodents and parasites like the flea, that in turn affected humans who came in contact with them. When viruses cross this species barrier humans initially have little natural resistance, and large numbers died. Humans in ancient times either survived or died as there were no treatments. Those humans who survived, however, developed immunity toward certain diseases. This survivor acquired immunity could be passed to offspring by antibodies in breast milk and from the mother's blood through the placenta to the unborn child, and this afforded protection to that generation from another outbreak of that infectious disease. This is often thought of as a reason a particular disease reoccurs in following generations. As populations increased other viruses such as mumps, rubella, and polio were seen in 9000 BCE as people settled in Egypt along the River Nile. Typically, it was a fertile flood plain bringing all sorts of microbes into contact with people and their animals. More food resulted in more people in which the microbes thrived and the viruses persisted as high concentrations of people were infected, all of whom were susceptible.

People moved throughout the Mediterranean basin during the next several thousand years and continued to domesticate wild animals including cattle, sheep, horses, goats, pigs, cats, and dogs. These animals brought infectious viruses with them from the wild. Zoonotic viral infections prevailed and included influenza and rabies transmission from animal to human. Some zoonotic viruses specific to species were not an immediate threat to humans, but were present nevertheless awaiting the opportunity. Diseases of ancient people evolved into what we see today in the modern age.

Early humans suffered colds, influenza, and diarrhea caused by viruses just as humans do today. Influenza viruses crossed the species barrier from waterfowl to pigs to humans. In ancient Egypt poliomyelitis, the polio virus infection, was present in an 18th dynasty Egyptian priest. The foot of his mummy showed evidence of deformity. Likewise, evidence of the virus was found in the mummy of the 19th dynasty ruler Siptah, 1197–1191 BCE. The smallpox virus devastated the ancient world as seen on the many mummies buried over 3000 years ago during the reign of Pharaoh Ramesses V (1149–1145 BCE). Mummies from thousands of years show evidence of smallpox.

In Classical Greece based on Athens Agora and Corinth data, life expectancy if one survived to 15 would be 37–41 years. Between 429 and 426 BCE there was a plague that affected Greece (Athens), Libya, Egypt, and Ethiopia thought to possibly be typhus, typhoid fever, or viral hemorrhagic fever, that killed 75,000–100,000. In 430 BCE the Athens army and many civilians died from a smallpox outbreak. A few years later, around 412 BCE, Northern Greece had an influenza epidemic. As the Romans expanded, they carried their microbes all through the area which was then part of the Roman Republic. Another smallpox outbreak occurred between 165 and 190 CE called the Antonine Plague. During this period in the Roman Empire, infections, possibly of smallpox and measles, affected roughly 25–33% of the Roman population killing 5–10 million people throughout the ancient Roman Empire which at that time included Britain, Europe, the Middle East, and Northern Africa; infection may have killed at least one Roman Emperor. Romans continued to spread the Empire and viral infections. The measles virus was also seen as extremely infectious in the ancient world and canine distemper virus and rinderpest virus were everywhere. It may be that they were first transmitted to humans from domesticated dogs or cattle. Another outbreak which may have been smallpox occurred in Europe between 250 and 266 CE and is known as the Plague of Cyprian.

Elsewhere in the world diseases also occurred. During the Han Dynasty in 217 CE in China was the Jian'an Plague, possibly of typhoid fever or viral hemorrhagic fever. In Japan around 700 CE the plant described to have yellowing leaves *Eupatorium lindleyanum* was often infected with what we now know as the tomato yellow leaf curl virus, and later between 735 and 737 CE there was a smallpox epidemic affecting 2 million people, which at that time was one-third of the Japanese population.

Outbreaks of smallpox, measles, influenza, and rabies continued for many years in early European towns and cities. During the late Roman Empire, they attempted to control the outbreaks of disease by advances in architecture, when in fact they may have been making matters worse. Rather than disposing of bad water they simply tossed it in the streets and the run-off would contaminate the water supply – leading to exposure to disease-causing organisms. Among the plethora of diseases that caused childhood death were measles, influenza, yellow fever, polio, and smallpox.

Crusades and Muslim conquests further spread infections of smallpox and measles causing epidemics in Europe in the 5th and 7th centuries. In the early Middle Ages (Europe, from the late 5th century or early 6th century to the 10th century CE) measles and other viruses spread throughout the highly populated countries of Europe and North Africa. During this time there was an increase in life expectancy of about 10 years (30–35 years).

More than 50 plagues occurred in England between 526 and 1087. Among these was an outbreak of rinderpest virus, a disease of cattle; it is closely related to the measles virus and has been documented since Roman times. The disease, which originated in Asia, was first brought to Europe by the invading Huns in 370. Later invasions of Mongols, led by Genghis Khan and his army, started pandemics in Europe in 1222, 1233, and 1238. The infection subsequently reached England following the importation of cattle from the continent.

The first confirmed outbreak was in August 1485 at the end of the Wars of the Roses, which has led to speculation that it may have been brought over from France by French mercenaries. In 1485 after the victory at the Battle of Bosworth the army suddenly went down with "the English sweat", which is considered to have been a viral infection caused by a medieval ancestor of the hantavirus. During the 15th and 16th centuries disease was prevalent among people who traveled from the continent and from greater distances, bringing diseases of all sorts. The medieval hantavirus outbreak might have originated in France where Henry VII had recruited soldiers for his army. Another epidemic hit London in the hot summer of 1508 where people died within a day. The streets were deserted apart from carts transporting bodies, and King Henry declared the city off limits except for physicians and apothecaries. By 1510 the influenza pandemic affected Asia, North Africa, and Europe. The disease spread to Europe, arriving in Hamburg in July 1529 where 1000–2000 victims died within the first few weeks.

Suddenly people were traveling more frequently all over and carried with them their microbes and discovered new microbes and diseases. During the 16th century, following travel to North America, smallpox and measles occurred among the Aztec, in Prussia, in Switzerland, and throughout northern Europe. Scientists today suggest that the disease was a combination of viral infections including species of influenza, poxvirus, measles, smallpox, and hantavirus. By the 1560s these viral infections including influenza had become pandemics affecting most of the explored world including Asia, Africa, Europe, and the Americas. Although attempts at limiting the outbreak were made, including restrictions on trade and travel, isolation of the stricken, the fumigation of buildings, and the killing of livestock, they did not work and disease spread across the lands. Medical references to influenza and these other infections date from the late 15th and early 16th centuries, but infections almost certainly occurred long before then. During an influenza epidemic that occurred in England between 1557 and 1559, thought to affect 5% of the population, 150,000 died from the infection. It should be noted that the mortality rate was nearly five times that of the 1918–1919 pandemic.

The first pandemic that was reliably recorded began in July 1580 and swept across Europe, Africa, and Asia. The mortality rate was high; 8,000 died in Rome. Early colonial America, 1616–1620, saw infections in southern New England, especially affecting the native Wampanoag people. Looking in retrospect and with the modern studies of virus, bacteria, and fungi, it is possible to determine that infections of leptospirosis with Weil syndrome, of yellow fever, bubonic plague, influenza, smallpox, chickenpox, typhus, hepatitis B, and hepatitis D occurred, with estimated deaths of 30–90%. One thousand deaths occurred due to a smallpox epidemic between 1633 and 1635 in Massachusetts Bay Colony and the 13 colonies of British North America. In 1634–1640 smallpox and an influenza epidemic affected the Wyandot people of North America and killed 15,000–25,000. There was an outbreak in 1648 of Central America yellow fever, and in 1677–1678 the Boston smallpox epidemic killed 750–1,000 people.

The next three measles and smallpox pandemics occurred in the 18th century, including 1781–1782, which was probably the most devastating in history, beginning in November of 1781 in China and reaching Moscow by December of that year. In February 1782 it hit Saint Petersburg, and by May it had reached Denmark. Within six weeks, 75% of the British population were infected and the pandemic soon spread to the Americas.

The Americas and Australia remained free of measles and smallpox until the arrival of European colonists between the 15th and 18th centuries. Along with measles and influenza, smallpox and yellow fever were taken to the Americas by the Spanish. Smallpox was endemic in Spain, having been introduced by the Moors from Africa. In 1519, an epidemic of smallpox broke out in the Aztec capital Tenochtitlan in Mexico which was believed to have been started by the army of Pánfilo de Narváez, who followed Cortés from Cuba, and had an African slave suffering from smallpox aboard his ship. When the Spanish conquests entered the capital in the summer of 1521, they saw it strewn with the bodies of smallpox victims. The epidemic, and those that followed during 1545–1548 and 1576–1581, eventually killed more than half of the native population. When the Europeans traveled to the New World during the time of the Spanish conquests they carried and spread viruses including smallpox, measles, yellow fever, and influenza and other disease to the indigenous people who had no natural resistance to the viruses, and millions died during epidemics. In the territory of New Spain in present day Mexico in the 16th century during the time of the Spanish conquests in 1519–1520 there was a Mexican smallpox epidemic killing 5–8 million which was 23–37% of the native population. The Cocoliztli epidemics of 1545–1576 are now known to be viral hemorrhagic fevers characterized by high fevers and bleeding and were caused by an indigenous viral agent and was aggravated by unusual climatic conditions. This Cocoliztli epidemic killed 5–15 million people or about 27–80% of the native Mexican population. Another viral outbreak of influenza occurred in 1580 and further decimated the indigenous population. The illnesses collectively were called Cocoliztli and were then a mysterious illness. If they survived the illness the people were generally immune, but could be carriers of the virus. Most of the Spanish were immune to smallpox having already been through multiple smallpox epidemics; thus, smallpox enabled an army of fewer than 900 men to defeat the Aztecs and conquer Mexico. Many Native American populations were devastated later by the inadvertent spread of diseases introduced by Europeans. In the 150 years that followed Columbus's arrival in 1492, the Native American population of North America was reduced by 80% from diseases, including measles, smallpox, and influenza. The damage done by these viruses significantly aided European attempts conquer the New World.

By the 18th century, smallpox was endemic in Europe. There were five epidemics in London between 1719 and 1746, and large outbreaks occurred in other major European cities. By the end of the century 400,000 Europeans were dying from the disease each year. Smallpox reached South Africa in 1713, having been carried by ships from India as shipping was a popular method for the movement of people, and their illnesses, at that time, and by 1789 the disease struck Australia. In the 19th century, smallpox became the single most important cause of death of the Australian Aborigines.

The first known cases of dengue fever occurred in Indonesia and Egypt in 1779. Trade ships brought the disease to the US, where an epidemic occurred in Philadelphia in 1780. There was an influenza pandemic in 1889–1890 that killed over one million people worldwide.

Meanwhile, newly emerging infectious diseases were posing an increasingly significant threat to human health. The majority are of zoonotic origin to which the human population was particularly suspectable. The increase and the intensification of animal farming may have exasperated this situation. The increase in the number of sheep, cows, pigs, fowl, and wild animals may have contributed. The Irish Great Famine of 1845–1852 was attributed to a disease in potatoes. The mold that was seen during the blight was actually a virus. The disease, called "curl", is caused by potato leafroll virus, and it was widespread in England in the 1770s, where it destroyed 75% of the potato crop. Another emerging disease was rabies, an often-fatal disease, caused by the infection of mammals with the rabies virus. Recently, in the 21st century, it has been relegated to a disease that affects wild mammals such as foxes and bats, but it has a long history and is one of the oldest known virus diseases: rabies is a Sanskrit word (rabhas) that dates from 3000 BCE, which means "madness" or "rage", and the disease has been known for over 4000 years. Descriptions of rabies can be found in Mesopotamian texts, and the ancient Greeks called it "lyssa" or "lytta", meaning "madness". References to rabies can be found in the Laws of Eshnunna, which date from 2300 BCE. Aristotle (384–322 BCE) wrote one of the earliest undisputed descriptions of the disease and how it was passed to humans. Celsus, in the 1st century CE, first recorded the symptoms called hydrophobia and suggested that the saliva of infected animals and humans contained a slime or poison. He invented the word "virus" to describe this contagion. Rabies does not cause epidemics, but the infection was greatly feared because of its terrible symptoms, which include insanity, hydrophobia, and death. Little was known about the cause of the disease until 1903 when Adelchi Negri (1876–1912) first saw microscopic lesions, now called Negri bodies, in the brains of rabid animals. Paul Remlinger (1871–1964) soon showed by filtration experiments that they were much smaller than protozoa, and even smaller than bacteria. Thirty years later, Negri bodies were shown to be accumulations of particles 100–150 nanometers long, now known to be the size of rhabdovirus particles, the virus that causes rabies. At the turn of the 20th century, evidence for the existence of viruses was obtained from experiments with filters that had pores too small for bacteria to pass through, and the term "filterable virus" was used to describe them.

This brings the story of disease to the modern era. It becomes a bit more intertwined with other events but the march continues, even to the current day. Until the 1930s most scientists believed that viruses were small bacteria, but following the invention of the electron microscope in 1931 they were shown to be completely different, to a degree that not all scientists were convinced they were anything other than accumulations of toxic proteins.

The situation changed radically when it was discovered that viruses contain genetic material, the building blocks of life, in the form of DNA or RNA. Once they were understood as distinct biological entities, they were soon shown to be the cause of numerous infections in humans, plants, animals, and even bacteria and fungi. Of the many diseases of humans that were found to be caused by viruses in the 20th century one, smallpox, has been eradicated. Diseases caused by viruses such as HIV, measles, and influenza have proved to be more difficult to control. Other diseases, such as those caused by arboviruses, have presented new challenges. As humans changed their behavior, so have viruses. In ancient times the human population was too small and isolated for large pandemics to occur and, in the case of some viruses, too small for them to survive. In 1900 the world average life span was 31–32 and a jump in human life spans occurred in 1950 with the discovery of antibiotics; the world average life span became 45.7–48, and in 2019–2020 the world average is 72.6–73.2 years..

It was not until the middle of the 20th century, when infant mortality was approximately 40–60% of the total mortality, that this situation improved due to better hospital care, antibiotics, better treatment for diseases, and better education of physicians. In the 20th and 21st centuries, increasing population densities, revolutionary changes in agriculture and farming methods, and high-speed travel all contributed to the spread of new viruses and the re-appearance of ancient ones. Like smallpox, some viral diseases might be conquered, but new ones have taken shape, such as severe acute respiratory syndrome (SARS); others will continue to emerge. To make matters more interesting, it is now possible to make new viruses. Attributes of different viruses can be combined with another, and "the beat goes on". The need to be able to detect and identify these "new viruses" becomes all the more important.

Let us take a brief view of the modern world. Human metapneumovirus, which is a cause of respiratory infections including pneumonia, was discovered in 2001. Smallpox virus was a major cause of death in the 20th century, killing about 300 million people. It has probably killed more humans than any other virus in history. In 1966 an agreement was reached by the World Health Assembly (the decision-making body of the World Health Organization) to start an "intensified Smallpox eradication program" and attempt to eradicate the disease within ten years. At the time, smallpox was still endemic in 31 countries including Brazil, the whole of the Indian sub-continent, Indonesia, and sub-Saharan Africa.

During the Second Boer War (1899–1902) measles was rife among the prisoners in the British concentration camps and accounted for thousands of deaths. Before the introduction of vaccination in the US in the 1960s there were more than 500,000 cases of measles each year resulting in about 400 deaths. In developed countries children were mainly infected between the ages of three and five years old, but in developing countries half of children were infected before the age of two. Measles remains a major problem in densely populated, less-developed countries with high birth rates and lacking effective vaccination campaigns. By the mid-1970s, following a mass vaccination program that was known as "make Measles a memory", the incidence of measles in the US had fallen by 90%. Humans are considered the only natural host of the measles virus. Immunity to the disease following an infection is lifelong.

Poliomyelitis treatment required the iron lung for some patients, as during the polio epidemic of 1960. Half of the exposed population had polio deformities. During the summers of the mid-20th century, parents in the US and Europe dreaded the annual appearance of poliomyelitis (or polio), which was commonly known as "infantile paralysis". The disease was rare at the beginning of the century, and worldwide resulted in only a few thousand cases per year. By the 1950s there were 60,000 cases each year in the US alone and an average of 2,300 in England and Wales. During 1916 and 1917 there had been a major epidemic in the US with more than 27,000 cases and 6,000 deaths recorded. There were 9,000 cases in New York City. At the time, nobody knew how the virus was spreading. Many New York City inhabitants, including scientists, thought that impoverished slum-dwelling immigrants were to blame even though the prevalence of the disease was higher in the more prosperous districts such as Staten Island, a pattern that had also been seen in cities like Philadelphia. Many other industrialized countries were affected at the same time. In particular, before the outbreaks in the US, large epidemics had occurred in Sweden. The reason for the rise of polio in industrialized countries in the 20th century has never been fully explained. The disease is caused by a virus that is passed from person to person by the fecal-oral route, and naturally infects only humans. Poor living conditions and sanitation in over-populated communities may have contributed to the outbreak. People started to improve sanitation and hygienic food preparation. Although the virus was discovered at the beginning of the 20th century, its ubiquitous nature was unrecognized until the 1950s. It is now known that fewer than 2% of individuals who are infected develop the disease, and most infections are mild. During epidemics the virus was effectively everywhere, which explains why public health officials were unable to isolate a source. The development of vaccines in the mid-1950s led to mass vaccination campaigns which took place in many countries. Some vaccines were unproperly tested and did not have good outcomes. Polio cases fell dramatically, however, after the vaccines were forced on the world; the last outbreak was in 1979. In 1988 the World Health Organization along with others launched the Global Polio Eradication Initiative, and by 1994 the Americas were declared to be free of the disease, followed by the Pacific region in 2000 and Europe in 2003. At the end of 2012, only 223 cases were reported by the World Health Organization. Mainly poliovirus type 1 infections occurred in undeveloped areas, 122 occurring in Nigeria, 1 in Chad, 58 in Pakistan, and 37 in Afghanistan.

The human immunodeficiency virus (HIV) is another virus that, when the infection is not treated, can cause acquired immunodeficiency syndrome (AIDS). Some virologists believe that HIV originated in sub-Saharan Africa during the 20th century and resulted in over 70 million individuals being infected by the virus. By 2011, an estimated 35 million had died from AIDS, making it one of the most destructive epidemics in recorded history. HIV-1 is one of the most significant viruses to have emerged in the last quarter of the 20th century.

When the influenza virus undergoes a genetic shift creating a new strain many humans have no immunity to the new strain, and if the population of susceptible individuals is high enough to maintain the chain of infection, pandemics occur. The genetic changes usually happen when different strains of the virus co-infect animals (zoonic), particularly birds and swine. Although many viruses of vertebrates are restricted to one species, influenza virus is an exception. The last pandemic of the 19th century occurred in 1899 and resulted in the deaths of 250,000 people in Europe. The virus, which originated in Russia or Asia, was the first to be rapidly spread by people on trains and steamships. A new strain of the virus emerged in 1918, and the subsequent pandemic of Spanish flu was one of the worst natural disasters in history. The death toll was enormous; 50 million people died from influenza worldwide. Reportedly 550,000 deaths were caused by the disease in the US, that is, ten times the US losses during the First World War, and 228,000 deaths in the UK. In India there were more than 20 million deaths, and in Western Samoa 22% of the population died. Although cases of influenza occurred every winter, there were only two other pandemics in the 20th century. In 1957 another new strain of the virus emerged and caused a pandemic of Asian flu; although the virus was not as virulent as the 1918 strain, over one million died worldwide. The next pandemic occurred when Hong Kong flu emerged in 1968, a new strain of the virus that replaced the 1957 strain. Affecting mainly the elderly, the 1968 pandemic was the least severe, but 33,800 were killed in the US.

New strains of influenza virus often originate in East Asia; in rural China the concentration of ducks, pigs, and humans in close proximity is the highest in the world. The most recent pandemic occurred in 2009, but none of the last three caused anything near the devastation seen in 1918. Exactly why the strain of influenza that emerged in 1918 was so devastating is a question that still remains unanswered.

Arboviruses are viruses transmitted to humans and other vertebrates by blood-sucking insects. These viruses are diverse; the term "arbovirus" which was derived from "arthropod-borne virus" is no longer used in formal taxonomy because many species of virus are known to be spread in this way. There are more than 500 species of arboviruses, but in the 1930s only 3 were known to cause disease in humans: yellow fever virus, dengue virus, and pappataci fever virus. More than 100 of such viruses are now known to cause human diseases including encephalitis. Yellow fever is the most notorious disease caused by a flavivirus. The last major epidemic in the US occurred in 1905, and during the building of the Panama Canal thousands of workers died from the disease. Yellow fever originated in Africa, and the virus was brought to the Americas on cargo ships by the *Aedes aegypti* mosquito that carries the virus. The first recorded epidemic in Africa occurred in Ghana, in West Africa, in 1926. In the 1930s the disease re-emerged in Brazil. Epidemics continue to occur in areas around the world. In 1986–1991 in West Africa, over 20,000 people were infected, 4,000 of whom died. In the 1930s, St. Louis encephalitis, eastern equine encephalitis, and western equine encephalitis emerged in the US. The virus that causes La Crosse encephalitis was discovered in the 1960s, and West Nile virus arrived in New York in 1999. As of 2010, the dengue virus is the most prevalent arbovirus and increasingly virulent strains of the virus have spread across Asia and the Americas along with yellow fever and other arboviruses.

Hepatitis, recognized since ancient times, is a disease of the liver. Symptoms include jaundice, a yellowing of the skin, eyes, and body fluids. There are numerous causes, including viruses – particularly hepatitis A virus, hepatitis B virus, and hepatitis C virus. Throughout history epidemics of jaundice have been reported, mainly affecting soldiers at war. This "campaign jaundice" was common in the Middle Ages. It occurred among Napoleon's armies and during most of the major conflicts of the 19th and 20th centuries, including the American Civil War, where over 40,000 cases and around 150 deaths were reported. The viruses that cause epidemic jaundice were not discovered until the middle of the 20th century. The names for epidemic jaundice, hepatitis A, and for blood-borne infectious jaundice, hepatitis B, were first used in 1947; following 1946 the two diseases were distinct. In the 1960s, the first virus that could cause hepatitis was discovered. This was hepatitis B virus, which was named after the disease it causes. Hepatitis A virus was discovered in 1974. The discovery of hepatitis B virus and the invention of tests to detect it have radically changed many medical and some cosmetic procedures. The screening of donated blood, which was introduced in the early 1970s, has dramatically reduced the transmission of the virus. Donations of human blood plasma and Factor VIII collected before 1975 often contained infectious levels of hepatitis B virus. Until the late 1960s, the un-hyenic use of hypodermic needles, which were often reused by medical professionals, and tattoo artists' needles were a common source of infection. It was not until the 1990s that needle exchange programs were established in Europe and the US to prevent the spread of infections by intravenous drug users. These measures also helped to reduce the subsequent impact of HIV and hepatitis C virus.

Epizootics are outbreaks (epidemics) of disease among animals that have continued since ancient times; it could even be said that they are the same ancient diseases. During the 20th century significant epizootics of viral diseases in animals, particularly livestock, occurred worldwide. The many diseases caused by viruses include foot-and-mouth disease, rinderpest of cattle, avian and swine influenza, swine fever, and bluetongue of sheep. Viral diseases of livestock can be devastating to both farmers and the wider community, such as the outbreak of foot-and-mouth disease in the UK in 2001. Rinderpest which first appeared in East Africa in 1891, a disease of cattle, spread rapidly across Africa. By 1892, 95% of cattle in East Africa had died. This resulted in a famine that devastated the farmers and nomadic people, some of whom were entirely dependent on their cattle. Two-thirds of the population of Maasai people died. The situation was made worse by epidemics of smallpox that followed in the wake of the famine. In the early years of the 20th century rinderpest was common in Asia and parts of Europe. The prevalence of the disease was steadily reduced during the century by control measures that included the vaccination of livestock and careful inventory control of imports and exports of livestock. By 1908 Europe was free from the disease. Outbreaks did occur following the Second World War, but these were quickly controlled. The prevalence of the disease increased in Asia, and in 1957 Thailand had to appeal for aid because so many buffaloes had died that the paddy fields could not be prepared for rice growing. Russia west of the Ural Mountains remained free from the disease – Lenin approved several laws for the control of the disease – but cattle in eastern Russia were constantly infected with rinderpest that originated in Mongolia and China where the prevalence remained high. India controlled the spread of the disease, which had retained a foothold in the southern states of Tamil Nadu and Kerala, throughout the 20th century, and had eradicated the disease by 1995. Africa suffered two major panzootics in the 1920s and 1980s. There was a severe outbreak in Somalia in 1928, and the disease was widespread in the country until 1953. In the 1980s, there were outbreaks in Tanzania and Kenya. Of the non-human animal viruses whiteflies (*Trialeurodes vaporariorum*) are the vector of cassava mosaic virus. Recurrence of the disease in 1997 was suppressed by an intensive vaccination campaign. By the end of the century rinderpest had been eradicated from most countries. A few pockets of infection remained in Ethiopia and Sudan, and in 1994 the Global Rinderpest Eradication Program was launched by the Food and Agriculture Organization (FAO) with the aim of global eradication by 2010. In May 2011, the FAO and the World Organization for Animal Health announced that "Rinderpest as a freely circulating viral disease has been eliminated from the world". Foot-and-mouth disease is a highly contagious infection caused by an aphthovirus, and is classified in the same family as polio virus. The virus has infected animals, mainly ungulates, in Africa since ancient times and was probably brought to the Americas in the 19th century by imported livestock. Foot-and-mouth disease is rarely fatal, but the economic losses incurred by outbreaks in sheep and cattle herds can be high. The last occurrence of the disease in the US was in 1929, but as recently as 2001, several large outbreaks occurred throughout the UK and thousands of animals were killed and burnt. The natural hosts of influenza viruses are swine and birds, although it has probably infected humans since antiquity. The virus can cause mild to severe epizootics in wild and domesticated animals. Many species of wild birds migrate, and this has spread influenza across the continents throughout the ages. The virus has evolved into numerous strains and continues to do so, posing an ever-present threat of reemergence. In the early years of the 21st century epizootics in livestock caused by viruses continue to have serious consequences. Bluetongue disease, a disease caused by an orbivirus outbreak, was seen in sheep in France in 2007. Until then the disease had been mainly confined to the Americas, Africa, southern Asia, and northern Australia, but it is now an emerging disease around the Mediterranean. During the 20th century, many ancient diseases of plants were found to be caused by viruses. These included maize streak and cassava mosaic disease. As with humans, when plants thrive in close proximity or are cultivated, so do their viruses. This can cause huge economic losses and human tragedies. In Jordan during the 1970s, where tomatoes and cucurbits (cucumbers, melons, and gourds) were extensively grown, entire fields were infected with viruses. Similarly, in Côte d'Ivoire, 30 different viruses infected crops such as legumes and vegetables. In Kenya cassava mosaic virus, maize streak virus, and groundnut viral diseases caused the loss of 70% of the crop. Cassava is the most abundant crop that is grown in eastern Africa, and it is a staple crop for more than 200 million people. It was introduced to Africa from South America and grows well in soils with poor fertility. The most important disease of cassava is caused by cassava mosaic virus, a geminivirus, which is transmitted between plants by whiteflies. The disease was first recorded in 1894, and outbreaks of the disease occurred in eastern Africa throughout the 20th century, often resulting in famine. Of known plant viruses in the 1920s the sugarbeet growers in the western US suffered huge economic loss caused by damage done to their crops by the leafhopper-transmitted beet curly top virus. In 1956, between 25 and 50% of the rice crop in Cuba and Venezuela was destroyed by rice hoja blanca virus. In 1958, it caused the loss of many rice fields in Colombia. Outbreaks recurred in 1981, which caused losses of up to 100%. In Ghana between 1936 and 1977, the mealybug-transmitted cacao swollen-shoot virus caused the loss of 162 million cacao trees, and additional trees were lost at the rate of 15 million each year. In 1948, in Kansas, US, 7% of the wheat crop was destroyed by wheat streak mosaic virus, spread by the wheat curl mite (*Aceria tulipae*). In the 1950s papaya ringspot virus – a potyvirus – caused a devastating loss of solo papaya crops on Oahu, Hawaii. Solo papaya had been introduced to the island in the previous century, but the disease had not been seen on the island before the 1940s. Such disasters occurred when human intervention caused ecological changes by the introduction of crops to new vectors and viruses. Cacao is native to South America and was introduced to West Africa in the late 19th century. In 1936, swollen root disease had been transmitted to plantations by mealybugs from indigenous trees. New habitats can trigger outbreaks of plant virus diseases. As of 1970, the rice yellow mottle virus was only found in the Kisumu district of Kenya, but following the irrigation of large areas of East Africa and extensive rice cultivation, the virus spread throughout East Africa. Human activity introduced plant viruses to native crops. The citrus tristeza virus (CTV) was introduced to South America from Africa between 1926 and 1930. At the same time, the aphid *Toxoptera citricidus* was carried from Asia to South America, and this accelerated the transmission of the virus. By 1950, more than six million citrus trees had been killed by the virus in São Paulo, Brazil. CTV and citrus trees probably coevolved for centuries in their original countries. The dispersal of CTV to other regions and its interaction with new citrus varieties resulted in devastating outbreaks of plant diseases. Because of the problems caused by the introduction, by humans, of plant viruses, many countries have strict importation and export controls on any materials that can harbor dangerous plant viruses or their insect vectors. Even without mutation, it is always possible that some meniscal, obscure parasitic organism may escape its accustomed ecological niche and be exposed to a new ecosystem with the dense populations that have become so conspicuous a feature of the earth, creating some fresh and perchance devastating mortality.

Emerging viruses are those that have only relatively recently infected the host species. In humans, many emerging viruses have come from other animals. We have talked about the viruses that jump to other species that cause disease in humans and are called zoonoses or zoonotic infections. In recent years, viruses are still here in the world as during ancient times, and they could be spread worldwide through every ecosystem, if we are not careful. In 2018 the Nipah virus outbreak in Kerala, India, caused by the Nipah virus infection had 17 fatalities. The Kivu Ebola epidemic 2018–2020 in the Democratic Republic of the Congo and Uganda had 2,280 deaths. In 2019 the measles outbreak in the Democratic Republic of the Congo had 7,018 casualties. In the 2019–2020 New Zealand measles outbreak and the 2019 Philippines measles outbreak, 415 died. In the 2019 Kuala Koh measles outbreak, there were 215 deaths, and in the Samoa measles outbreak, there were 83 casualties. In 2019–2020 a dengue fever epidemic affected the Asia-Pacific and Latin America with 3,930 fatalities. The Nigeria Lassa Fever epidemic, from 2019 to the present, has killed 247. In 2020, the Democratic Republic of the Congo had an Ebola outbreak killing 55. Between 2020 and the present, Nigeria's yellow fever epidemic has caused 296 deaths.

West Nile virus, a flavivirus, was first identified in 1937 when it was found in the blood of a feverish woman. The virus, which is carried by mosquitoes and birds, caused outbreaks of infection in North Africa and the Middle East in the 1950s, and by the 1960s horses in Europe fell victim. The largest outbreak in humans occurred in 1974 in Cape Province, South Africa, and 10,000 people became ill. An increasing frequency of epidemics and epizootics (in horses) began in 1996, around the Mediterranean basin, and by 1999 the virus had reached New York City. In the United States, mosquitoes carry the highest amounts of virus in late summer, and the number of cases of the disease increases in mid-July to early September. When the weather becomes colder, the mosquitoes die and the risk of disease decreases. In Europe, many outbreaks have occurred. In 2000 a surveillance program began in the UK to monitor the incidence of the virus in humans, dead birds, mosquitoes, and horses. The mosquito (*Culex modestus*) that can carry the virus breeds on the marshes of north Kent. This mosquito species was not previously thought to be present in the UK, but it is widespread in southern Europe where it carries West Nile virus. In 1997 an outbreak of respiratory disease occurred in Malaysian farmers and their pigs. More than 265 cases of encephalitis, of which 105 were fatal, were recorded. A new paramyxovirus was discovered in a victim's brain; it was named Nipah virus, after the village where he had lived. The infection was caused by a virus from fruit bats, after their colony had been disrupted by deforestation. The bats had moved to trees nearer the pig farm and the pigs caught the virus from their droppings. Several highly lethal viral pathogens are members of the Filoviridae. Filoviruses are filament-like viruses that cause viral hemorrhagic fever, and include the Ebola and Marburg viruses. The Marburg virus attracted widespread press attention in April 2005 after an outbreak in Angola. Beginning in October 2004 and continuing into 2005, there were 252 cases including 227 deaths. The Ebola virus epidemic in West Africa, which began in 2013, is the most devastating since the West Nile virus.

The COVID-19 pandemic and its ever-changing variants are affecting populations worldwide from 2019 to present; known as the coronavirus disease 2019, the COVID-19 SARS-CoV-2 virus has killed an estimated 5.5–22 million+. Severe acute respiratory syndrome (SARS) is caused by a new type of coronavirus. Other coronaviruses were known to cause mild infections in humans, so the virulence and rapid spread of this novel virus strain caused alarm among health professionals as well as public concern. Vaccines have been developed in several forms, not to be used on everyone. As for the coronavirus vaccine, the makers have tried to delay the release of the side-effects for 75 years; apparently this was done for the smallpox and polio vaccines. This often leads to misinformation about the pandemic and the constant use of masks and gloves. The exact origin of the SARS virus is not known, but SARS evidence suggests that it came from bats or was related to bats. The coronavirus that emerged in Wuhan, China, in November 2019, and spread rapidly around the world may have been a man-made modified bat virus. Subsequently named severe acute respiratory syndrome coronavirus 2 (SARC2), infections with the virus caused a pandemic with a case fatality rate of around 2% in healthy people under the age of 50, up to around 15% in those aged over 80. The fatality rate lowered but the infections increased because the new mutated virus was more contagious. Control measures were limited in part by fear, misinformation, prejudice, and stigmatization of infected people. Limited use of available virus detection methods limited the knowledge of where the virus was located in the environment, who was infected, and who had the virus and were not infectious. Unprecedented restrictions, in peacetime, were placed on international travel and curfews imposed in major cities worldwide with limited effect, because, in part, they did not know who was carrying the virus. One contagious person traveling with hundreds of healthy people could infect the whole group. Governments were not prepared for the scale of the pandemic worldwide; virology and epidemiology experts were complacent with regards to the efficiency of existing testing and monitoring systems and the use of vaccines. With these limitations it was difficult to control.

The discovery of the abundance of viruses and their overwhelming presence in many ecosystems has led modern virologists to reconsider their role in the biosphere. More than a trillion viruses likely exist on Earth, most being bacteriophages, and most are in the oceans. Microorganisms constitute more than 90% of the biomass in the sea. It has been estimated that viruses kill approximately 20% of this biomass each day and that there are 15 times as many viruses in the oceans as there are bacteria and archaea. Viruses are the main agents responsible for the rapid destruction of harmful algal blooms, which often kill other marine life, and help maintain the ecological balance of different species of marine blue-green algae, and thus adequate oxygen production for life on Earth.

The Human Genome Project has revealed the presence of numerous viral DNA sequences scattered throughout the human genome. These sequences make up around 8% of human DNA, and appear to be the remains of ancient retrovirus infections of human ancestors. These pieces of DNA have firmly established themselves in human DNA. Most of this DNA is no longer functional.

Viruses have transferred important genes to plants. About 10% of all photosynthesis uses the products of genes that have been transferred to plants from blue-green algae by viruses. We may not be able to see viruses with the naked eye, but through extensive viral research and understanding of their phylogenetic relationships we know they have existed everywhere humans have since prehistory.

Modern science has the advantage of not only seeing viruses and physically counting them, but understanding their biological relationships due to the discovery of DNA and RNA. Having these biomolecular tools has allowed science to understand phylogenetic relationships. This grouping of viruses by genetic relationship and indeed the type of disease that they cause has set the stage for virus detection. But, first, let us look at a few basics: what is a virus and how are they classified?

complacent with regard to [illegible] systems and the use of vaccines [illegible]

The discovery of the [illegible] in many ecosystems [illegible] the pleosphere [illegible]

[illegible]

2

Microbes, Fungi, Bacteria, and Viruses

The three main types of microbes, fungi, bacteria, and viruses, are discussed along with the various methods and techniques used to detect and classify them. This process is a short history of the evolution of detection from fungi, that are easily seen, to bacteria, that require a microscope, to viruses, that require an electron microscope. Groups were classified by those means as they became available. The discovery of genetics, DNA, and RNA and the sequencing of the genome made understanding of the phylogenetic relationships and the classification of organisms by their genetic relationships possible. The number of viruses sequenced has increased (Figure 2.1). The evolution of technology, such as computers and the mass spectrometer, have allowed the ability to detect peptides and more importantly those unique peptides associated with a particular strain of microbe. It is now possible to detect a microbe (fungi, bacteria, or virus) or many microbes of different types in a single sample using software.

Since they were first discovered scientists have wanted to be able to identify and classify the many different microbes, in particular bacteria, fungi, and viruses. It was an exciting time of discovery during the development of the disciplines of microbiology, mycology, and virology. The results were that in general an academic specialty was offered in the three fields. Sometimes bacteriology and virology were offered in the same academic area referred to as microbiology. One reason for this early separation was their size. Fungi were large multi-micron- to millimeter-sized organisms, bacteria were 0.5 to 2.0 microns, and viruses were nanometer sized; generally three orders of magnitude separated the fungi, bacteria, and viruses (Figure 2.2). Many methods were developed for the detection and identification of these microbes, and although some methods have many desired characteristics, none of them satisfy all the criteria and none can detect and identify all three types of microbes in a single sample preparation.

For many years this was the status – three groups of scientists, three disciplines of research, and various naming schemes. This status all changed with the application of molecular biology and the ability to determine

DOI: 10.1201/9781003106623-2

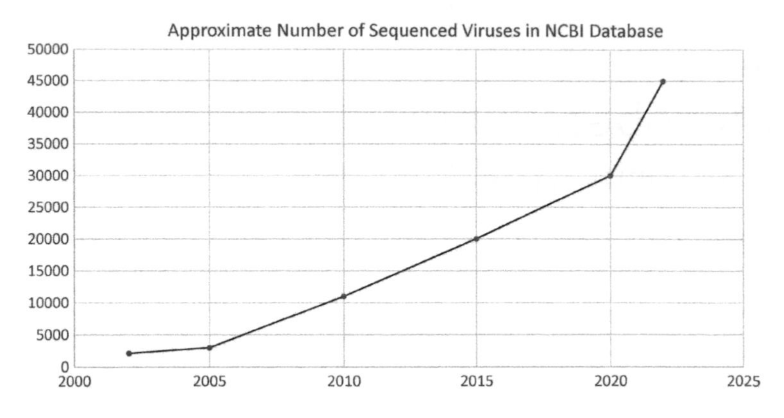

Figure 2.1 Growth of NCBI database of sequenced viruses.

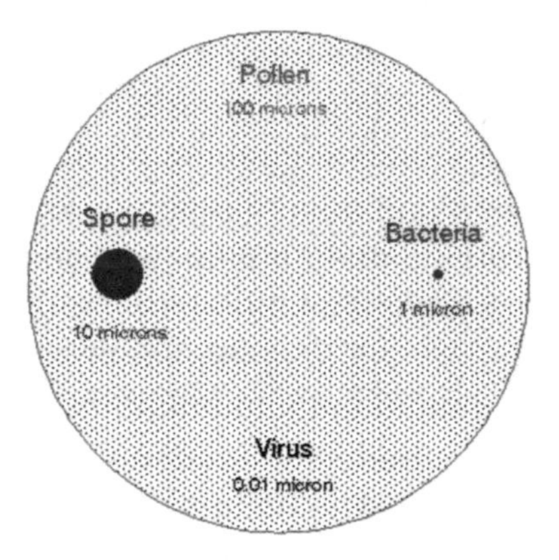

Figure 2.2 Relative size of pollen, fungi spores, bacteria, and virus. Pollen – 100 μm. Fungi – 10 μm. Bacteria – 1 μm. Virus – 0.01 μm.

microbes' genomic sequence. Standardization occurred, and order was established among the microbes. This change swept through all three disciplines in a few years. Classification schemes were changed to move those microbes closely related genetically into the phylogenetic mapping scheme. All types of microbes were moved around within the old schemes to create a new classification scheme. There were changes within the disciplines as a result; scientists re-learned the new schemes and saw some of their microbes listed and published with new names along with the old name in parentheses.

The prospect of needing a rapid single detection platform method presents many challenges. Some challenges are unique to epidemics, and others are common for all testing situations (Klietman and Ruoff 2001). Detection platforms should be capable of rapidly detecting and confirming hostile microbes, including modified or previously uncharacterized ones, directly from complex matrix samples, with no invalid results. Instruments should be portable, user-friendly, and capable of testing for multiple microbes simultaneously.

2.1 Fungi

2.1.1 What is a fungus?

Fungi (single: fungus) are eukaryotic organisms that include molds, yeasts, and mushrooms. The fungi are classified as a kingdom (Figure 2.3).

2.1.2 How are fungi detected/classified?

Fungi are known to most people as the mushrooms that we see and are separated from each other by eye. This early separation into groups by descriptive taxonomy means that even in ancient times the fungi were classified. Some are good to eat, and some are not. The discovery of micron-sized fungi came along later as technology enabled the discipline to expand, and these microbes were then included in the classification schemes, but mainly their outward characteristics were used to name the different fungal species. Those fungal species that affected people in ways other than being toxic were generally few, and their study was usually a specialized field within the overall medical microbiology discipline. However, those that

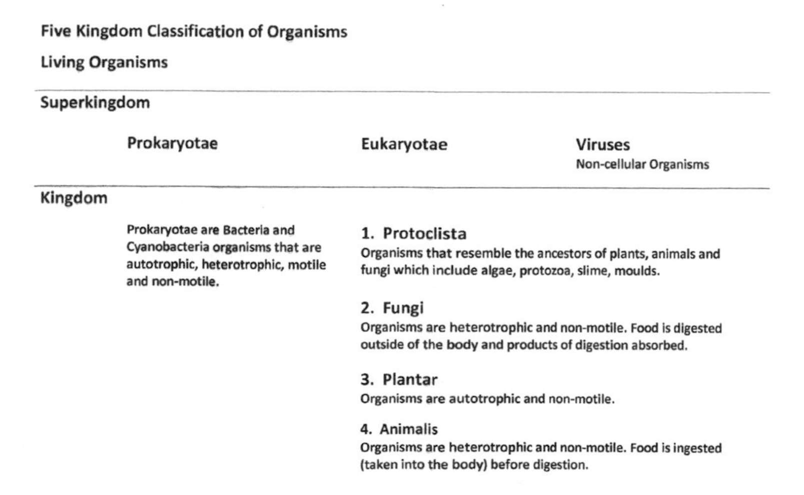

Figure 2.3 Classification of living organisms; notice the viruses are a dotted line as they are sometimes not considered to be living organisms.

affected plants, trees, and food sources were also important, and the field of mycology evolved among these disciplines. The discovery of genetics and the ability to genetically separate fungi further allowed them to be related by their phylogenic relationships.

2.2 Bacteria

2.2.1 What are bacteria?

Bacteria (singular bacterium) are ubiquitous microorganisms consisting of one biological cell. They are smaller than fungi but larger than viruses (Figure 2.2). Bacteria are classified in the superkingdom, Prokaryotae (Figure 2.3), and are known as prokaryotic microorganisms because they do not have nuclei that are partitioned by an intracellular membrane. Bacteria inhabit the soil, water, hot springs, and most habitats. Many bacteria have not been sequenced and are unclassified. The academic discipline is known as bacteriology.

2.2.2 How are bacteria detected and classified?

Early efforts used to classify and identify bacteria concentrated on growing them and learning what they metabolized and other physical features, such as color, edge of colony characteristics, if they were round or rods, and if they had an ability to stain with iodine. These differences are used to classify them into groups for naming purposes. Gram positive and Gram negative organisms could help identify infectious diseases and gave professionals the ability to identify the same microorganism – Gram negatives and Gram positives Worldwide, this ability to identify and classify bacteria quickly helped in the control of infections and outbreaks of infections associated with historical epidemics. The discovery of antibiotics for controlling bacterial infections produced a rapid increase in scientific work, and the discipline of microbiology grew along these lines of research.

The common and historical bacteriological methods have provided a basis for identification and frequently are still the routine identification methods. These techniques are generally based on the determination of the morphology, differential staining, and physiology of a bacterial isolate (sample). These tests can be performed by means of miniaturized and automated substrates that utilize screening methods to classify and identify the isolate. Popular systems in this classification and identification category are the VITEK (BioMerieux, Hazelwood, MO) and MicroLog (BioLog, Hayward, CA) that utilize metabolizable substrates and carbon sources or susceptibility to antimicrobial agents. Such systems have been used to identify microbes such as *Bacillus anthracis* (Baille et al. 1995), *Yersinia* spp. (Linde et al. 1999), and other pathogens (Odumeru et al. 1999). In addition, BioLog introduced a "dangerous pathogen" database to its MicroLog system. Although such techniques are still considered standard practice for the isolation and identification of infectious microbes, they are time-consuming, and often can take days to obtain even preliminary results.

The classification and identification methods that have resulted from the genomic methods have resulted in phylogenetic classification and the use of new methods. Such methods as mass spectrometry proteomics (MSP) can classify and identify sequenced bacteria using software. More than 437,192 prokaryotes have been sequenced.

2.3 Viruses

2.3.1 What are viruses?

Viruses (singular: virus) are considered to be infectious microbes that consist of a segment of nucleic acid (either DNA or RNA) surrounded by a protein coat. They are not included among the living cell-based microbes (Figure 2.3) but have an extensive classification due to their physical characteristics and their phylogenetic relationships (Figure 2.4). Viruses cannot replicate on their own but utilize the cellular apparatus to make copies and in this manner are very different from organisms in other kingdoms.

2.3.2 How are viruses detected and classified?

The discovery of viruses resulted in the development of new methods and improvements in all detection technologies: electron microscopy, molecular methods, direct detection, and mass spectrometer means. Improvements in both scanning and transmission electron microscopes improved the visualization of viruses. Resulting discoveries found multitudes of different surface features on viruses, the different sizes of viruses, and other features of viruses that resulted in their early characterization. Names of viruses were frequently associated with the host and type of illness. Viruses are frequently characterized by their wide variety of shapes and features, such as with envelope, without envelope, spherical, helical, and the bacterial phages which have head and tail arrangements. Viruses range from 20 to 350 nanometers (nm). Viruses are ubiquitous. Animals, insects, plants, and bacteria all have their own special viruses.

Attempts to bring order to the characterization of viruses were initially chaotic. The naming of new viruses is sometimes challenging. Putting order to

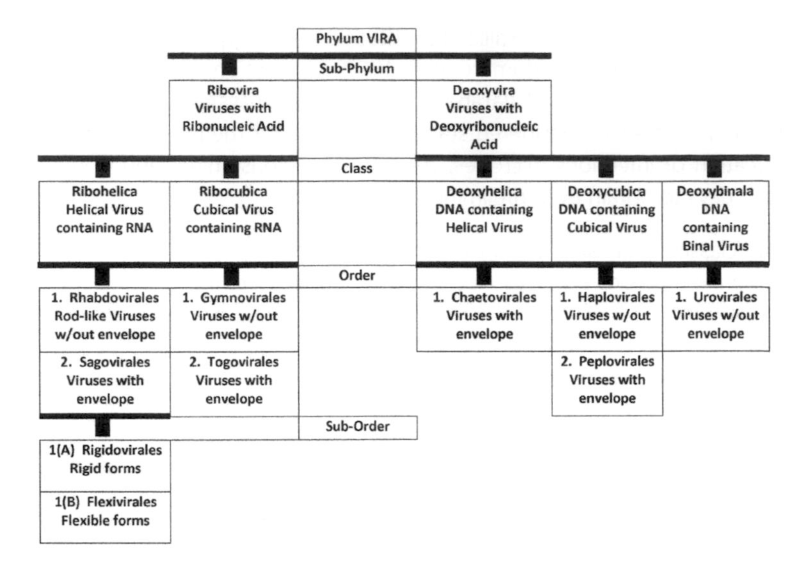

Figure 2.4 Classification of viruses according to whether they contain RNA or DNA.

this chaos is the responsibility of the International Committee on Taxonomy of Viruses (ICTV). Viruses are sorted (classified) by their phenotypic characteristics, such as their morphology, the nucleic acid type (DNA, RNA), how they replicate, the host organisms they inhabit, and what disease they are associated with. This classification scheme is an ongoing process; however, Figure 2.3 indicates where viruses are associated with other organisms; they are represented as a superkingdom because they are different from living things because they lack a cellular structure. Nevertheless, Figure 2.4 represents a refinement of this classification by dividing the viruses into two groups, those with RNA and those with DNA Figure 2.5 Further physical features divide the viruses according to those viruses with an envelope and those without an envelope. It is beyond the scope of this book to delve into the controversies of the naming and classification schemes; that is the function of the ICTV.

Since its discovery in the 1930s, electron microscopy has been widely used to look at all aspects of the nano world. Its use in visualizing viruses has allowed the discovery of their sizes and a plethora of features; morphology, surface proteins, and organization are just a few examples. Because it is looking at a virus it does not depend upon genomic information; this is a physical process. Genomic information is not needed to detect a new virus. Likewise, electron microscopy does not identify a virus.

DNA Viruses		
Double-stranded	Double-stranded	Single-stranded
With Envelope	Without Envelope	Without Envelope
Hepadnaviruses Herpesviruses Poxviruses	Adenoviruses Papillomaviruses Polyomaviruses	Parvoviruses

RNA Viruses				
+ RNA	+ RNA	- RNA	+/- RNA	+ RNA via DNA
Without Envelope	With Envelope	With Envelope	Double Capsid	With Envelope
Caliciviruses Picomaviruses	Coronaviruses Flaviviruses Togaviruses	Arenaviruses Bunyaviruses Filoviruses Orthomyxoviruses Paramyxoviruses Rhabdoviruses	Reoviruses	Retroviruses

Figure 2.5 Division of the (a) DNA viruses and the (b) RNA viruses based on the presence or absence of an envelope.

Molecular methods are often considered to be polymerase chain reaction (PCR) based and similar methods based upon the genomic information of viruses. Other methods, though not related, are the immunoassay means where the detection of virus depends upon the detection of the viral antibodies being detected.

Nucleic acid–based methods usually combine PCR amplification with the simultaneous detection of amplified products based on changes in reporter fluorescence. For specific detection, the change in fluorescence relies on the use of dual-labeled fluorogenic probes. An increase in fluorescence indicates that the probe has hybridized to the target DNA, and this principle is used for a variety of tests that rely on the quantitative presence or absence of targeted sequences. However, the main PCR format used for biothreat agents is usually specific target detection, and a wide variety of primer and probe combinations are available from many companies in a multitude of configurations. Many of these specific target configurations rely on mechanistic variants in the primer and probe construction and combinations which can include TaqMan probes and other primers (Westin et al. 2001). They are available commercially and can be customized. Recently, several companies have started to offer PCR kits in various formats for the detection of viruses. These kits simplify primer/probe design and facilitate remote and rapid detection and monitoring programs. Many of the nucleic acid approaches for the detection of biothreat agents are described in review articles (Ivnitski et al. 2003). Additional approaches investigate methods to detect nucleic acids from viruses using isothermal amplification or directly from samples without using an amplification step.

To increase the specificity of detection methods, immunoassays gained popularity in the 1990s despite the fact that they can test for only one analyte per assay. This limitation means that multiple simultaneous or sequential assays must be performed to detect more than one analyte in a sample or specimen. Advances in assay design and in matrix format have resulted in the development of multiplex assays that can be performed on multiple samples simultaneously by automated systems. However, the specificity of immunoassays is limited by antibody quality and sensitivity (detection limits ~10^5 cfu) which is typically lower than with PCR and other DNA-based assays. With improvements in antibody quality and assay parameters, it may be possible to increase immunoassay sensitivity and specificity in the future. Many different immunoassay formats are currently commercially available for a wide variety of detection needs. Many formats are similar to, or derived from, the classic sandwich assay based on the enzyme-linked immunosorbent assay (ELISA) design (Murray et al. 2003).

The current limitations of existing bioassays have become evident. They cannot generally identify unknown microbes. Identification is predicated on existing knowledge of a virus. Specific antibodies or specific primer/probe pairs are required. Many attempts have been made to extend this capability, but it will always come up short when confronted with the magnitude of the problem; there is always one more virus. Current techniques are expensive and frequently require special skill, expensive preparation, manufacturing, and reagents. Something else is indicated.

The invention of the Integrated Virus Detection System (IVDS) in the late 1990s is discussed in Chapter 6. IVDS was invented to detect intact viruses and be able to separate them according to size and concentration by simply counting them. This capability allowed for the rapid screening of samples without the encumbering reagents and complications of molecular biology methods. A highly reliable method for the detection of viruses emerged – rapid screening by IVDS and confirmation by other methods such as immunoassay, PCR, or MSP. Fast detection followed by relatively fast methods to identify followed by comprehensive characterization (MSP). This method is ideally suited for the detection of un-characterized, or for practical purposes unknown, viruses, since it is a physical detection method and not dependent on genomic information.

Chapter 7 discusses the MSP technology that has been rapidly developing during the last 25 years, mass spectrometry proteomics approaches to microbe identification. Early MS devices were able to detect the byproducts produced by microbes, such as fatty acids, and developed work arounds for detecting viruses, but it was not until improvements in the mass spectrometry techniques coupled with proteomic approaches that an acceptable next evolution in the detection and identification of microbes was achieved; individual peptides could be detected and identified. Associating these peptides using the ABOid software allowed the sorting of those peptides unique to a particular sequenced virus which allowed for accurate identification.

Mass spectrometry proteomic methods gather a wide range of information about microbes. The techniques are not limited by reagents or prior knowledge of a microbe. This approach is not a directed or targeted approach to identification. Often tens of thousands of unique peptides are associated with each particular sequenced virus (more than 50,000 viruses).

Initially mass spectrometry (MS) methods used profiles of both pyrolytic products and fatty acids as specific microbial biomarkers for identification purposes. For example, the commercial microbial identification system (MIDI Inc., Newark, DE) continues to use gas chromatography (GC) of cellular fatty acid methyl esters for the identification of bacteria. This method has been used to identify and differentiate *Bacillus* spores and other potential biological warfare agents (BWAs). Furthermore, MIDI Inc. introduced the Sherlock Bioterrorism Library that can be added to its identification system to specifically target biothreat agents and other organisms of interest. The MIDI Sherlock system containing the MIDI BIOTER database has been awarded AOAC Official Methods of Analysis status for the confirmatory identification of *Bacillus anthracis* (AOAC International 2004). On the other side, the Chemical Biological (CB) MS system developed by DOD used the pyrolytic processing of biothreat agents to generate mainly fatty acid methyl ester-products analyzed by MS-based methods for discrimination.

Advancements in genomics, hardware, and software have made this new capability possible. A rapid increase in the number of sequenced microbes, bacteria, fungi, and viruses has occurred in the last few years, and this trend is expected to continue. In addition, mass spectrometers have improved during the last several years, and in some cases, what was dreamed possible only a few years ago is now taking place in our laboratories. Very fast acquisition (thousands of peptides in minutes) and high resolution (100,000 at m/z 400) have resulted in sensitivities (subfmole = 10^{-16}–10^{-17} mole by LC/MS) that are continuing to improve. Computer capabilities have improved rapidly in the last few years, making possible calculations that took longer than a week to perform a couple of years ago in less than a few minutes.

All this capability is rapidly improving at such a rate that it is not unusual to see in an active laboratory often several generations of various types of equipment used to evaluate microbes. Changes are taking place as research is taking place; results from previous research re frequently improved by new equipment. Frequently, these improvements are seen in what was once limited by hardware or computers in both sensitivity and the time to analyze.

The convergence of genomic sequencing, mass spectrometer advancements, faster computers, and software has made it possible to sequence and analyze peptides using the mass spectrometer. Software can be used to calculate all the known peptides for a sequenced microbe and then sort out all the microbes from each other to determine those peptides that are unique to a given fungus, bacteria, or virus. It is then simple to construct a phylogenic tree to identify an organism or organisms and relate them to their near neighbors. This is a genetically accurate and sensitive method for detecting and identifying all the microbes in a complex mixture.

It sounds simple enough, pick a mass spectrometer, select a fast computer, and utilize appropriate software, and you suddenly can detect and identify microbes. Scientists and engineers have been busy, and there are many types of mass spectrometers and several types of computers and many sorts of software. Advancements historically have proceeded over a wide front. It seems that advancements have come at a fast pace for MS hardware, computers, and software.

MSP can detect and classify all three types of microbes (fungi, bacteria, and virus). The technology has been tested, and many double-blind trials have proven its capability. It has high accuracy similar to or better than common PCR means. The results also demonstrate that MSP can detect and identify microbes that are not sequenced (not in the genetic databases) to the level that can be matched by their unique peptides (family, genus, species) based upon their genetic similarities, for example some species of *E. coli* are sequenced to species and others are not. MSP can detect and identify a genus of *E. coli* for those species not sequenced. In this manner, microbes can be classified according to the degree of match to their taxonomic hierarchy making the detection and grouping of unknown microbes scientifically sound.

When MSP methods are combined with other methods such as the IVDS, greater capabilities are realized. In the IVDS/MS method many samples can quickly be analyzed for the presence of viruses and only the positive samples continue to the MS for identification. This combination also improves the sensitivity of detection by concentrating and separating the viruses from complex mixtures. Such hybrids improve the discrimination of negative and positive samples and improve overall identification time.

Different types of mass spectrometers have been developed; their operating principles are discussed, as well as a brief examination on how they are used. Since there are different types of mass spectrometers it is not surprising that they have in general different preparation methods. The analysis of the MS files then is determined by what sort of information can be obtained.

It is clear that a combined microbe detection platform must be capable of detecting a variety of microbes in complex samples. This capability is vital because samples may contain toxins, fungi, bacteria, viruses, and other types of analytes. In some instances, microbes may have been deliberately altered through genetic, antigenic, or chemical modifications or may represent new or uncommon variants of known microorganisms. Such modifications can make detection difficult using the common methods. Therefore, the only way to overcome these problems in a timely fashion would be the rapid sequencing of nucleic acids or deducing nucleotide sequence information from amino acid sequences of proteins. The latter approach has the advantage of including protein toxins, which can be easily modified to escape detection by any method that does not rely on amino acid sequence information.

Developments in the area of MS allow for the application of this analytical platform for the analysis of nucleic acids and proteins and obtaining sequence-based information about the microbial world. This information is suitable for the detection, classification, and reliable identification of all microbes. Three important considerations are the sensitivity, specificity, and reproducibility of such a platform.

An important consideration in virus is the collection and handling of samples. Airborne and waterborne samples generally must be concentrated from large volumes to detect low levels of target analytes. In many cases airborne samples must also be transferred to a liquid because most detection platforms process only liquid samples. The efficiency of recovery from concentration and extraction procedures can vary and affects detection limits. It is advantageous to isolate or concentrate target analytes prior to analysis in complex sample matrices such as powders or food.

In general, detectors that use nucleic acid detection systems such as PCR are more sensitive than antibody-based methods. PCR requires a clean sample and is unable to detect protein toxins and other non-nucleic acid-containing analytes such as prions. Furthermore, cultures of the target organism are not available for archiving and additional tests after the PCR analysis is complete. MSP techniques can identify thousands of specific viruses and archive the results to be analyzed again if new viruses are discovered to check if the new virus was in fact contained in the historical sample.

Specificity is as important as sensitivity in the detection of biothreat agents. High specificity is important to minimize background signals and false positive results from samples that are often complex, uncharacterized mixtures of organic and inorganic materials. Specificity can be affected not only by background particles, but also by high concentrations of competing antigens and DNA. The high sensitivity of PCR, for example, can also be a major weakness because contaminating or carry over DNA can be amplified, resulting in false-positive results. MS methods that use sequencing information for detection and identification purposes are usually characterized by high specificity that is limited only by the sequencing information available in databases.

Being able to replicate the method is as important as sensitivity and specificity. Reproducibility is an important requirement for detection platforms because systems that do not provide reproducible results are unreliable and may exacerbate a terrorist event. Many factors can affect the repeatability of bioassays, including the stability and consistency of reagents and differences in assay conditions. These variations can often be reduced by standardizing assay conditions and procedures; however, the best solution would be the use of one of the methods that do not use reagents.

3 Indirect Methods of Detecting Viruses

3.1 Introduction

What did people do before 1930? They could not see a virus; they knew about bacteria as they were seen for the first time with the invention of optical microscopes and the resultant discovery of bacteria first seen by Antonie Van Leeuwenhoek in 1676. Much research followed regarding the microworld and especially the unicelled organisms discovered. Among these discoveries was the ability to filter the microorganisms. Filters were used to "clean" up the samples and remove the microorganism. Smaller particles in the sample capable of causing infections were discovered. One of the noteworthy examples of this discovery was the tobacco mosaic virus (TMV). It was observed that tobacco plants were developing spots or "mosaic"-like patterns on the leaves. It was later determined that TMV actually infects a wide range of other plants. TMV is often considered the first "virus" to be discovered and has been known since the 19th century. It was a non-bacterial infectious disease and was known to be smaller than a bacterium because it could pass through filters designed to remove bacteria. The filtered liquid was found to still infect tobacco plants. It was not until the invention of electron microscopes in the 1930s that it was visualized and classified accordingly. It should be noted that filters were developed in the 1990s that were marketed as virus-free filters. This was later proven incorrect as some viruses passed through the membranes. The mechanics of how this is done are still a mystery.

There were several indirect observations of non-bacterial infections. Some thought the culprits were small bacteria, but after a long period of discovery and the presence of the new electron microscopes, it was determined that there was a new group of microbes, and they were called viruses.

DOI: 10.1201/9781003106623-3

4
Electron Microscopy

4.1 Introduction

Historically, the next major invention for the detection and discovery of viruses was the electron microscope. It was invented by physicists in the 1930s after the discovery of electrons that would behave as light particles and thus could be manipulated like light by magnetic means. These magnetic lenses became the apertures and condensers of this new type of microscope. The resolution was greater than light microscopes and was such that they could visualize particles in the nanometer size, and this opened up a whole new world for research and discovery. This began the discovery of many sorts of viruses and their classification by physical features and the diseases they cause.

Electron microscopes evolved from the study of physics by Max Knoll and Ernst Ruska producing the first transmission electron microscope (TEM). The first commercial TEMs were made in 1939 and the 1940s. These instruments and their capability to visualize the microworld have blossomed into a tool for science spanning many disciplines. Ruska was awarded the Nobel Prize in physics in 1986.

Electron microscopy has been divided into two main groups with variations within each group –TEM and scanning electron microscopy (SEM). These two groups are ideal to visualize individual viruses. This technology has been able to determine the size of a virus, from the small polio virus (~24 nm) to the larger smallpox virus (300 nm). This enables the grouping of viruses by size (Figure 4.1). Round virus, rod shaped viruses, odd shaped viruses, environmental viruses, and viruses found in every niche in the biological world have been seen and studied. From the TEM images, the first drawing of a virus was made showing the surface features and shapes. Further, it has been possible to determine the physical features associations with cells. SEM can render a 3D image of samples with a resolution of a few nanometers. This ability has provided diagrams of details of the surface structures of viruses.

It is important to note that electron microscopes (both TEM and SEM) are routinely used to study nearly everything from electric circuits in

DOI: 10.1201/9781003106623-4

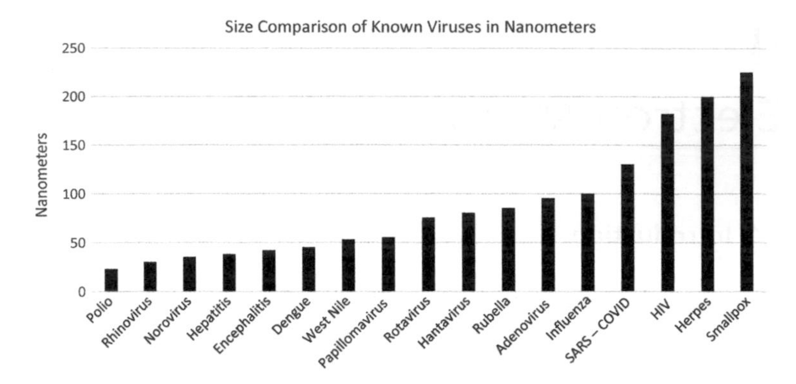

Figure 4.1 The size in nanometers (nm) of common viruses as determined by the electron microscope.

microprocessors, to animal cells, plant cells, bacteriophages, fungi, bacteria, and viruses. Details of cells such as plasma membranes and the cell nucleus and nanotubules can be seen and studied. Electron microscopy is used in almost every scientific discipline.

4.2 Transmission electron microscopy

TEM is a method involving electrons which have a wavelength approximately 100,000 times shorter than photons of visible light. A TEM has a greater resolving power than a photonic microscope. Specimens are thinly sliced (100 nm thick) and small. The image results from the interactions of the electrons passing through the thinly sliced sample. Using magnetic lenses, the image can be magnified and focused to make an image that can be directed to a fluorescent screen, photographic film, or a display.

4.2.1 How does TEM work?

TEMs are capable of imaging at a significantly higher resolution than light microscopes, owing to the smaller wavelength of electrons. This enables the instrument to capture fine detail – even as small as a single column of atoms, which is nearly 100,000 times smaller than a light microscope. Transmission electron microscopy is a major analytical method in the physical, chemical, and biological sciences. TEM instruments have multiple operating modes including conventional imaging, scanning TEM imaging (STEM), diffraction, spectroscopy, and combinations of these. Even within conventional imaging, there are many fundamentally different ways that contrast is produced, called image contrast mechanisms.

Contrast can arise from position-to-position differences in the thickness or density (mass-thickness contrast), atomic number (Z contrast), crystal structure, or orientation (crystallographic contrast or diffraction contrast). The slight quantum-mechanical phase shifts that individual atoms produce in electrons that pass through them (phase contrast) are used, as well as the effects of the energy lost by electrons on passing through the sample (spectrum imaging) and more. Each mechanism is used to gather different kinds of information to produce an image which depends not only on the contrast mechanism but on how the microscope is used. All of the settings of lenses, apertures, and detectors influence the quality of the image. TEM are capable of extraordinary nanometer- and atomic-resolution information, which, in ideal cases, can reveal not only where all the atoms are but what kinds of atoms and how they are bonded to each other. TEMs are regarded as an essential tool for nanoscience in both the biological and materials fields.

4.2.2 How do you use electron microscopy?

Foremost, the use of an electron microscope implies careful observation by a skilled operator. Many adjustments and tuning of the instrument are required for optimal results. Likewise, there is considerable skill required for preparing the specimen before using the instrument.

TEM uses an ultrathin specimen to reveal the internal structure of a sample. The specimen is often placed in epoxy resin. Thin sections (100 nm) are sliced, often with a diamond knife using an ultramicrotome. The thin slices are placed on a small grid, stained to improve contrast, and then viewed on the TEM; the steps are summarized:

- Collect specimen (clean up)
- Cut to size suitable for processing in ultramicrotome
- Embed in epoxy
- Specimens are thinly sectioned using an ultramicrotome
- Put thin sections onto a grid
- Stained
- Viewed using a TEM
- Record images (photographs or digital)

Sample preparation methods vary widely, depending on the nature of the sample. Sometimes hundreds of methods are tested to select the proper procedure for a given specimen. The process is routine once the methods are determined.

4.2.3 How do you identify a new virus?

Because electron microscopy is similar to using a regular photonic microscope, identifying a new virus is easy. You simply look for it. The preparation may require fine turning and the source might need careful preparation, but a new virus is simply viewed the regular way. New viruses that are unknown require some careful observation to ascertain the virus details, such as size and features. It should be noted that aside from careful observation and possible fine tuning of the sample preparation, adding a new virus or rather the discovery of a new virus is made without reagents or new materials (Figure 4.2).

4.3 Scanning electron microscopy (SEM)

SEMs use electrons to produce images of the surface. Electrons interact with atoms on the surface of a sample and produce images by integrating the various signals about topography and composition using information obtained by electron beam position and intensity of the detected signal. Some SEMs have a resolution better than 1 nm.

Examples of SEMs are given in Figure 4.3. Notice the nearly portable model. These small desk-top models allow for the SEM examination of samples nearly anywhere. They are simple to use and give great results.

4.3.1 How does scanning electron microscopy work?

Images are the result of a highly manipulated or focused electron beam which scans the surface to be studied. As the beam interacts with the sample being studied it is influenced by the features of the sample. The beam electrons are reflected and gathered by a detector, or detectors depending on the configuration of the SEM. The resulting black and white images reveal details of less than 1 nanometer and resulting magnifications between 5 and 500,000. The intensity of this interaction between the electron beam and the surface of the specimen depends on the nature of the specimen. The surface features have slightly different effects on the electron beam, depending on how the SEM is configured, with either one or two active detectors, and an intensity is measured. This creates a gray-scale image to which artificial color can be digitally assigned (Figure 4.4).

4.3.2 How do you use scanning electron microscopy?

This is a similar process as for TEM, the difference is the specimen is not sliced into thin layers. All that process is eliminated. The specimen is attached to a specimen holder to be placed in the SEM. Often, before going

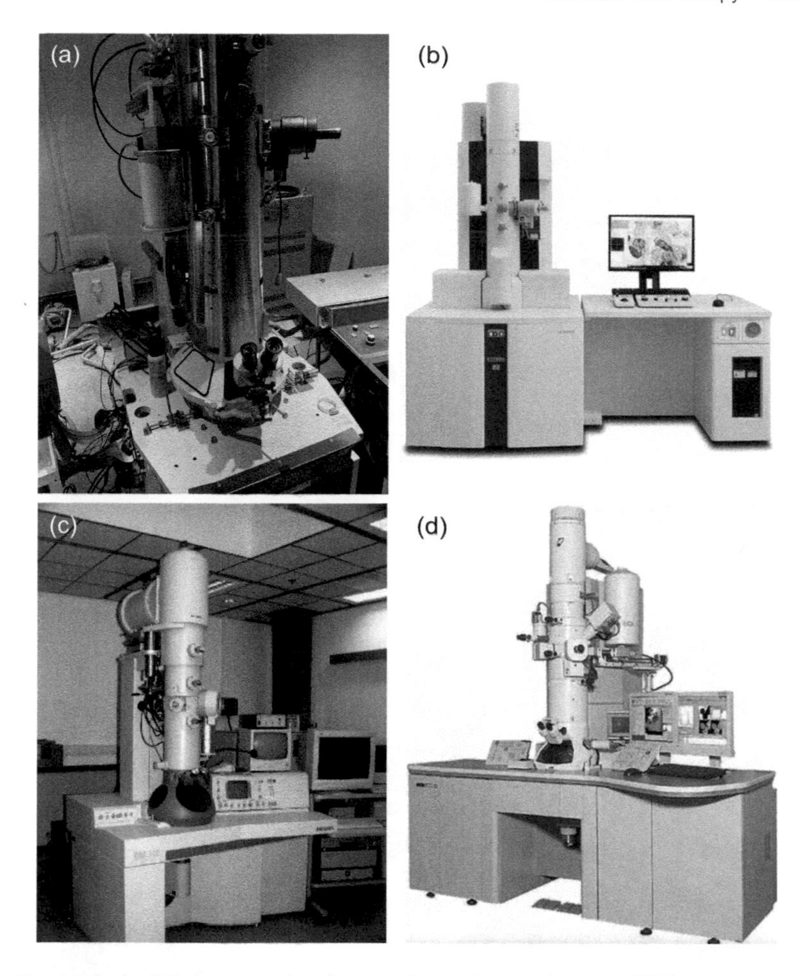

Figure 4.2 (a–d) Various examples of a transmission electron microscope (TEM).

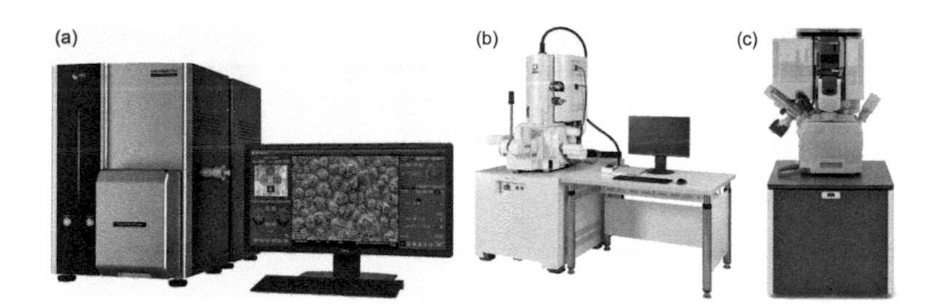

Figure 4.3 (a) SNE-4500M Tabletop SEM, (b) JEOL JSM-IT800HL, and (c) Thermo Scientific Helios 5 Dual Beam SEM.

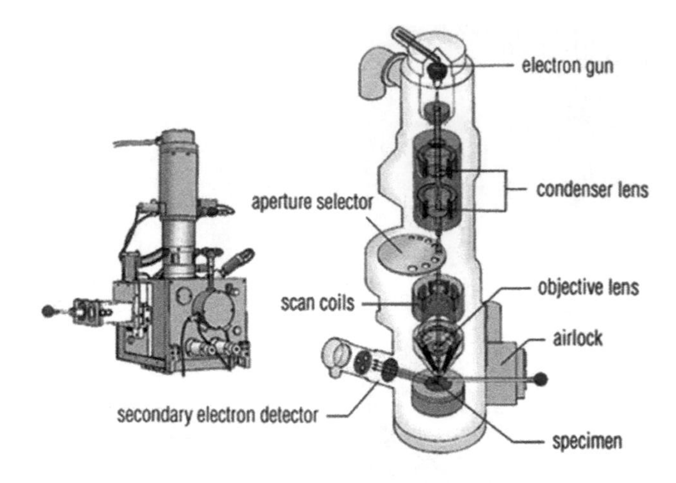

Figure 4.4 Working diagram of SEM.

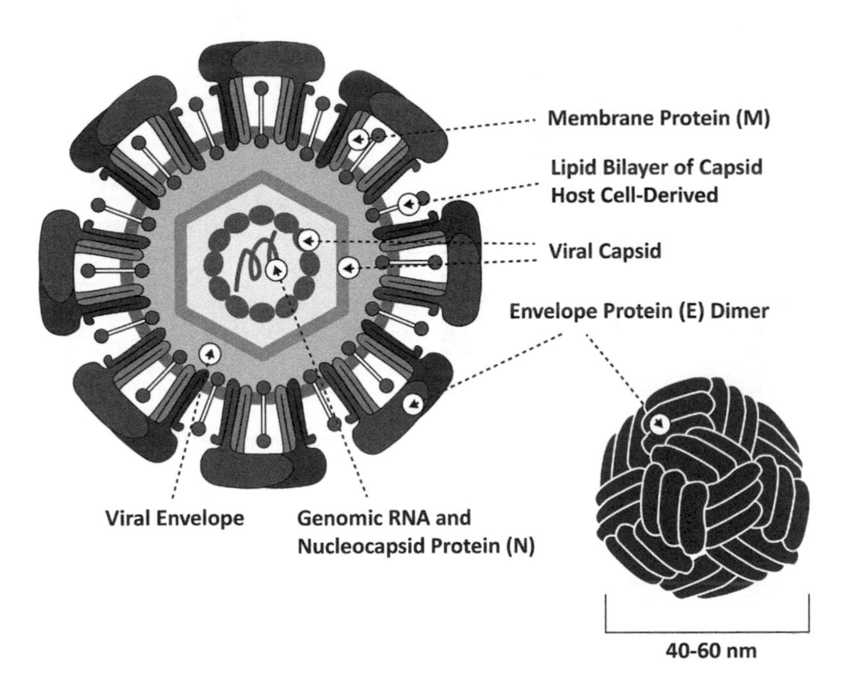

Figure 4.5 Illustration of dengue fever virus showing features (approximately 49 nm).

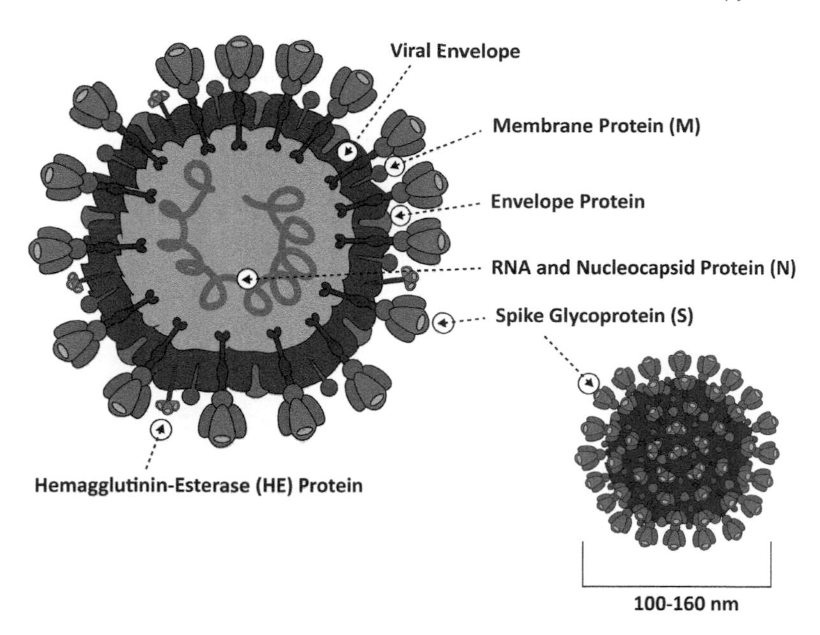

Figure 4.6 Coronavirus (COVID-19) illustration (approximately 130 nm) showing features.

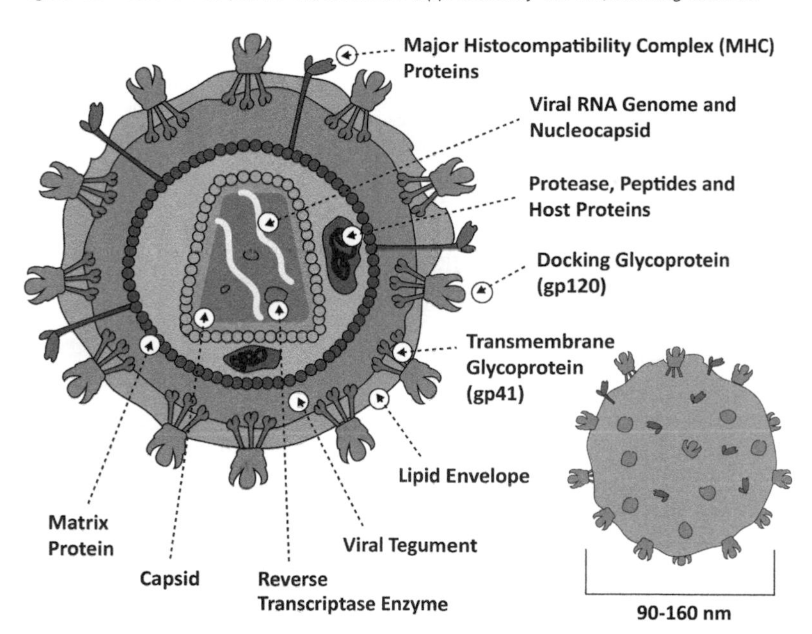

Figure 4.7 Illustration of human immunodeficiency virus (HIV) (approximately 180 nm) with features.

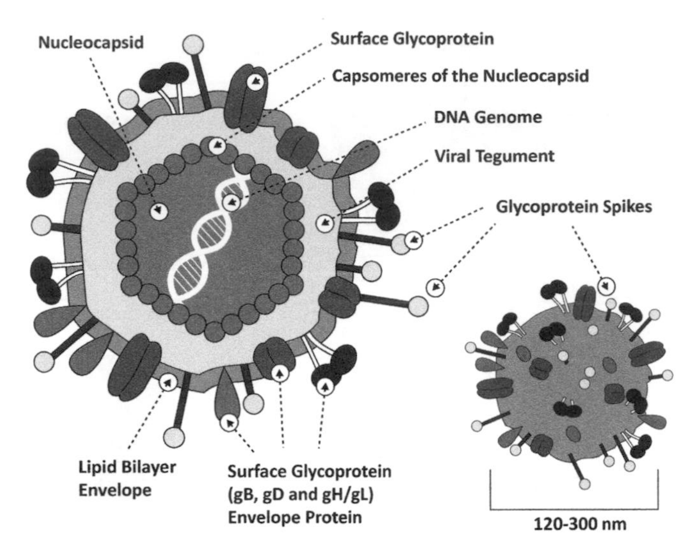

Figure 4.8 Illustration of herpes simplex virus (HSV) (approximately 200 nm) with features. There are several strains of the herpes virus that range from 120 to 300 nm.

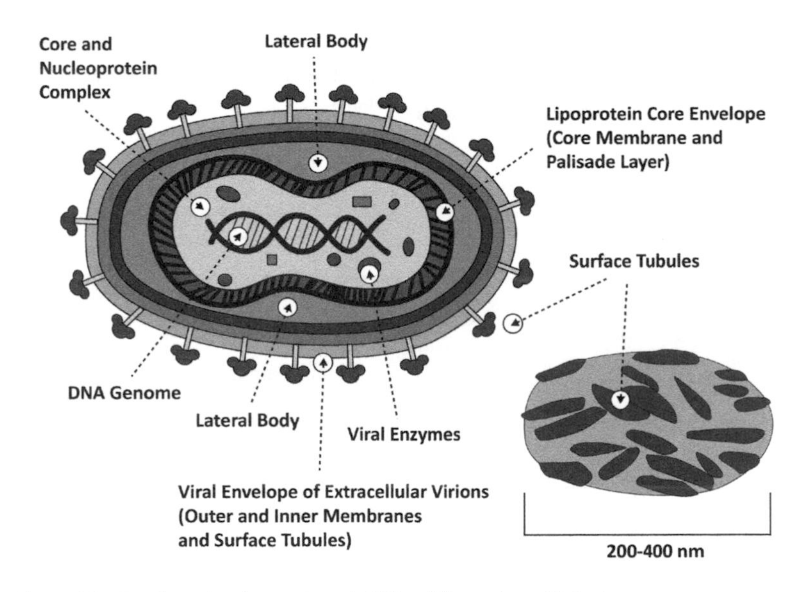

Figure 4.9 Smallpox virus (approximately 230 nm) illustration with features.

into the instrument, the sample is coated with metal, usually gold, to render better contrast. The coated specimen in placed in the SEM and images are recorded. To summarize:

- Collect specimen
- Coat specimen with metal which is used to improve contrast
- Place in SEM
- View specimen
- Record images (photographs or digital)

4.3.3 How do you identify a new virus?

Much in the same way as the TEM, the scientist uses careful observation. Sometimes special processing of the sample, coating, and handling are needed for the best photographs or digital images. Seeing something new requires skill and patience, but the new virus does not require special reagents or handling. The best magnification depends on the size of the virus.

4.4 Illustrations of viruses based on TEM and SEM visualizations

TEM and SEM microscopes have allowed details of viruses to be illustrated in structural diagrams. Research has filled in the functional features and enhanced illustrations which have been useful and rewarding. Figures 4.5–4.9 are illustrations of viruses seen in Figure 4.1.

into the instrument, the sample [illegible]
[illegible]
[illegible] to such notice.

- [illegible]
- [illegible]
- [illegible]

[illegible]

5
Molecular Methods for Detecting Viruses

5.1 Introduction

What are molecular methods? This is a good question to ask anyone that is seeking to know about how to classify and identify living things. Let us begin by first discussing DNA, RNA, and the genomic approach to understanding organisms. Prior to the discovery of DNA and the fact that this molecule contains all the information necessary to describe an organism, most living things were classified according to their description. One example of this is the classification of fungi. Prior to the 1970s fungi were organized and classified according to descriptive taxonomy. All mushrooms with certain features were classified together and given a Latin name that corresponded. If a mushroom had a red top, it might be named *Fistulina hepatica* meaning little red top mushroom. This classification scheme was known as the Friesian System of Descriptive Taxonomy; it was named after Dr. Elias Fries in 1758 who was a Swedish botanist and developer of the first system that was used to classify fungi.

The discovery of DNA and the resulting explosion of information that resulted gave a means to classify organisms according to their genetic relationship to each other. Fungi were reorganized according to genomic taxonomy. The kingdom Fungi has gained several new members on the basis of molecular phylogenetic analysis. This in turn has resulted in the Fungal Tree of Life (AFTOL) Project funded by the US National Science Foundation. AFTOL exists to discuss the uncertainties that remain about the exact relationships of many fungi and where they belong in the phylogenetic tree where organisms are classified by how close they are related generically.

It can be seen that fungi have gone through a rapid change in classification during the last couple of decades. The result is a classification method based on DNA which allows for all close relatives to be grouped together. New discoveries can be added to the phylogenetic tree in a systematic fashion. If this was complicated for the fungi which people have been seeing and eating for thousands of years, imagine the difficulty in classifying viruses

DOI: 10.1201/9781003106623-5

which were first visualized in the 1930s with the discovery of the electron microscope.

Molecular methods for classifying viruses took the same path as molecular methods used to classify fungi. The difference is that some viruses have been classified by their DNA and others by their RNA. Also, some viruses are so closely related that it is sometimes difficult to separate them into different families and genus, let alone a separate species. This classification is further complicated by the ease with which viruses mutate. This has become such a complex issue that many virus species have been categorized into sub-species. There is much academic discussion about where to place some of the new emerging viruses. Attention has been focused in the main on those viruses of interest to humans and animals.

The number of sequenced viruses has increased from a few viruses in the early 2000s to nearly 50,000 viruses in 20 years as illustrated in Figure 2.1. This rapid increase has advantages and some disadvantages. The main advantage is that we can classify the new viruses and place them in their appropriate place on the phylogenetic tree. One major challenge is keeping up with the identification of a new virus by genomic methods. The new viruses are a challenge for some new detectors simply because of the rapidly increasing and large numbers. Molecular-based virus detectors are prioritized according to demand. This is the result of the time it takes to develop a test protocol and the costs. The detection and identification of an obscure bacteriophage is less important than the detection and identification of a threatening new virus.

Discussed in other chapters are methods for detecting viruses that are not limited by the constraints of molecular biology. Examples are electron microscopy methods that visualize viruses, the IVDS method for detecting and counting viruses, as well as mass spectrometry proteomics (MSP) which uses software; all of these methods have largely solved some of these constraints of molecular biology.

The primary molecular biology methods used for detecting viruses are polymerase chain reaction (PCR) and antibody methods. In the following sections of Chapter 5 these methods and how they work will be discussed, and we will answer the question – how do you add new viruses?

5.2 Polymerase chain reaction (PCR)

The PCR method is a laboratory technique that takes hours to days to reach an analysis, although advances have been made that have resulted in small single use devices that take only minutes. The purpose of PCR testing is to find small amounts of DNA or RNA in a sample. These bits of genetic material are then increased using a process known as amplification to provide enough DNA or RNA to produce a detection and identification stage for a virus based on the genetic information.

Detecting viruses with RNA requires a slight modification of the historical PCR methods. This new technique is called reverse transcription-polymerase chain reaction (RT-PCR). RT-PCR uses RNA as starting material rather than DNA for in vitro nucleic acid amplification. Reverse transcriptase catalyzes DNA synthesis using RNA as the template. This makes the method an RNA-dependent DNA polymerase reaction. A middle product is produced called complementary DNA (cDNA). This is important because cDNA is more stable than RNA. As a result, RNA analysis in the clinical laboratory is nearly as rapid and sensitive as PCR DNA amplification and is commonly used in the diagnosis and quantification of RNA virus infections. Since many viruses contain RNA, this process has provided an important step forward in detection.

Unlike most bacteria, viruses are difficult to culture. Since RT-PCR allows the identification of viruses which are difficult to culture this is a useful capability and in particular for pathogens such as HIV (AIDS) and herpes. Advances in the technology and the application of various methods of RT-PCR and related techniques mean that analysis can be done in "real time", which means that a visible result is produced almost immediately. Historically, results were only visible at the end of the reaction and, as a result, RT-PCR and PCR are widely used diagnostic tools for detecting virus pathogens.

PCR revolutionized the study of DNA and is often considered an important advancement in molecular biology, to such an extent that its creator, Kary B. Mullis, was awarded the Nobel Prize in 1993.

The primary steps of PCR testing are:

1. Determine the genomic sequence of the target organism.
2. Select the portion of the sequence to represent that organism and isolate it from the genome, usually a piece of the DNA or RNA. This portion is snipped out of the long strand of base pairs that make up the genome. These "snippets" are then used to create the "primers". A primer is unique for a specific genome.
3. Primers are purified, prepared, and used in PCR instruments.
4. PCR testing.
5. Results.

The first step in this complex biochemical manipulation is to select the segment or "snippets" of a particular genome to be targeted for identification. This segment is that part of the genome that is to be amplified by PCR and is used as the basis for identification. The major part of this process is to first sequence the DNA or RNA of the organism to be tested. Some of these strings of DNA and RNA can be long and have thousands or tens of thousands of base pairs. "Primers" (which contain the "snippets") are purified and prepared and used in PCR instruments.

The concept is to isolate those portions of the string of DNA that are unique, usually by using specific software to sort the sequence. The unique partial genetic sequence is then used to represent the genus to which the organism belongs, e.g., influenza (*Alphainfluenzavirus).* A second partial sequence is usually selected that is unique to the family or sometimes the strain, e.g., influenza A or B. Sometimes a partial sequence is determined for a unique component of a microbe, e.g., the spike protein of the coronavirus (COVID-19). These steps are frequently complex and require expertise and skill to accomplish.

The first step is to design and create a "primer", the unique genetic segment. As mentioned, two segments of a particular genome or more are "selected" for processing where one segment represents the genus of the organism, and the second segment the strain of the organism or other unique features of the virus. Since the genetic sequences of thousands of organisms are available from the National Institutes of Health, a particular organism is selected and the genome examined for segments of their DNA or RNA for unique properties (base pair combinations) related only to that particular organism. Twenty to thirty base pairs are usually desired in selecting a segment for processing; any less and there are possible errors, and any more creates increases in processing time and other errors. This process can be challenging as there are often many unique segments. The two segments or more are identified by various means, usually by software for further research. These segments are then individually isolated, separated from the main genome, DNA or RNA, purified, and used as the "primer". The various primers are then used with various reagents in the many PCR instruments. The results are an identification of the organism based on their genome.

It is important in making these primers and the resulting probes to use the appropriate melting temperature (T_m); this is important in binding to the target "snippet" or sequence. The number of primers available has grown into the thousands.

PCR is influenced by many factors which include the primer, the "snippet" length, interactions with interfering factors such as metals, the occurrence self-folding during the assembly of the base pairs, melting temperature, annealing temperature, and GC (the nucleotides, guanine and cytosine) content of the "snippet".

The usual steps in PCR are frequently automated in a PCR instrument and are considered important for accurate and reliable operation and results.

- Primers' melting temperature (T_m) – ideally 62°C, but can range between 60 and 64°C. This is important for the PCR enzyme function.
- Annealing temperature (T_a) – depends on the length and composition of the primers, usually no more than 5°C below the T_m of the primers. Errors can result if this temperature is much lower or too high, leading to nonspecific amplifications or inefficient annealing.
- GC content – generally ideal at 50%, but can range between 35 and 65%.

To summarize the PCR process after the sample is introduced to the PCR instrument, what happens in the instrument?

- The sample is heated to denature the DNA (separating the DNA into two pieces).
- An enzyme called *Taq* polymerase is used (it builds two new strands of DNA, and this method is used to create a duplication of the original DNA).
- Repeat process in the thermocycler, to create copies (process is repeated 30–40 times to produce up to a billion exact copies of the DNA segment; this results in exponential amplification).
- Reagents are consumed and reaction stops (end point, plateau phase).
- Read via optical means (fluorescents) the color change produced by the amplified copies. PCR reactions contains a fluorescent reporter molecule (green dye) to monitor the accumulation of PCR product. As the product increases so does the fluorescence.
- Determine positive results using a computer.

Different PCR processes take measurement at the exponential phase, the liner phase, or the plateau phase depending on the type of instrument (Figure 5.1). Each has advantages and limitations. The plateau phase is where traditional PCR takes its measurements, also known as end-point detection; this can also produce variable results.

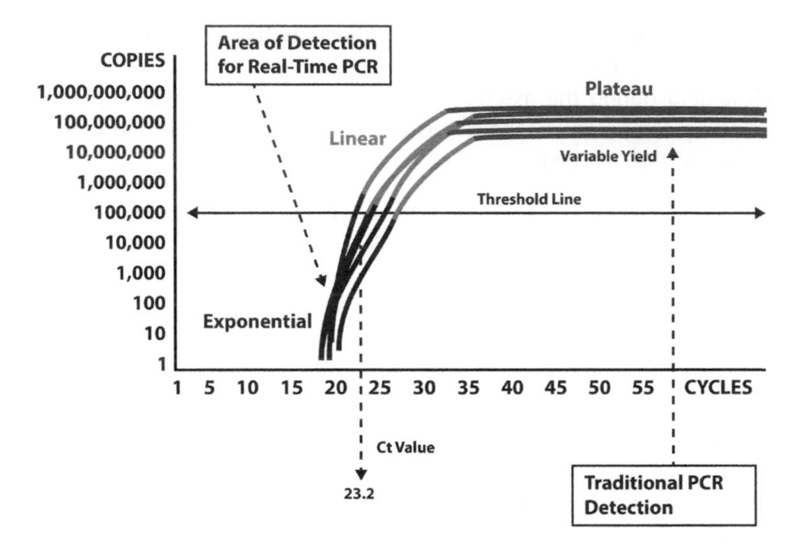

Figure 5.1 PCR phases. Traditional PCR measures at the plateau.

Figure 5.1 illustrates multiple samples which had the same amount of DNA in the beginning of the reaction and different quantities of PCR product by the plateau phase of the reaction due to variations in reaction kinetics. As a result, it would be more precise to take measurements during the exponential phase, where the replicate samples are amplifying exponentially.

Real-time PCR focuses on the exponential phase because it provides the most precise and accurate data for quantitation. Within the exponential phase, the real-time PCR instrument calculates two values. The threshold line is the level of detection at which a reaction reaches a fluorescent intensity above background. The PCR cycle at which the sample reaches this level is called the cycle threshold (C_t). The C_t value is often used to determine an estimate of the concentration by comparing the values of samples of unknown concentration with a series of standards.

Digital PCR counts individual molecules for absolute quantification. Digital PCR works by digitizing a sample into many individual real-time PCR reactions; some portions of these micro reactions contain the target molecule (positive) while others do not (negative). Following PCR analysis, the fraction of negative answers is used to generate an absolute answer for the exact number of target molecules in the sample, without reference to standards or endogenous controls.

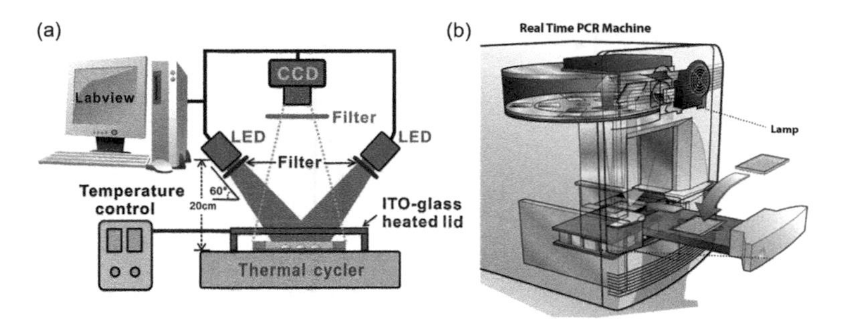

Figure 5.2 Illustrated view of a PCR instrument showing locations and features of components.

Figure 5.2 illustrates basic features of a PCR instrument. The different components are exposed, such as the thermal cycler, LEDs, the CCD camera, and the computer. Operation of the instrument is usually automated.

Figure 5.3 illustrates the typical output for a single sample and for the analysis of multiple samples and standards (Multiplex). In this manner multiple samples can be analyzed at the same time.

5.2.1 How do you add new viruses?

Adding a new virus requires a few steps. The virus needs to be identified and isolated. It then needs to be sequenced. Identifying the genomic material is next, followed by biomolecular processing to produce a useable primer ("snippet"). Once the primer is processed much of the PCR process is the same. This process takes time and skill (Figure 5.4a–d).

5.2.1.1 Examples of PCR instruments

Testing for COV-19 is used as an example of using PCR for the detection of a virus. The steps are outlined and explained.

The steps in Figure 5.5 start with: (1) take a sample from the patient's nose or mouth. Samples are taken from these locations because they are easily accessible and are known to contain the COVID-19 virus if present; (2) isolate the COVID-19 RNA from the patient sample. RNA extraction is a common laboratory procedure that can be performed with commercially available kits containing the appropriate methods used to separate RNA. Often a centrifuge is used to separate materials by their densities, and this step is important in the RNA extraction; (3) convert RNA into complementary DNA (cDNA) by an enzyme called reverse transcriptase. The RNA is converted into cDNA because the final step can only be performed on DNA.

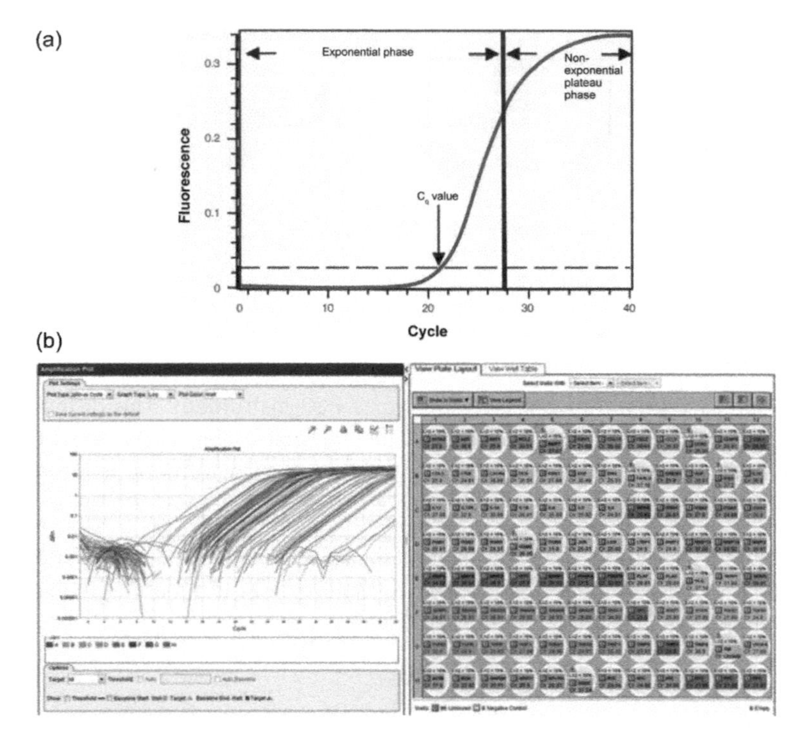

Figure 5.3 Example of PCR results for both (a) simple, single analysis and (b) multiplex analysis using several primers.

The cDNA is now combined with primers specific for COVID-19, a special enzyme called *Taq* polymerase and a special fluorescent probe. Primers are assigned to specifically test for COVID-19, based on information from the COVID-19 genome. All of these components are loaded into a plate and run in a real-time PCR machine and analyzed. Because these primers are specific to COVID-19, the PCR test will not amplify any cDNA if the original sample does not contain COVID-19 RNA. In this manner, when the

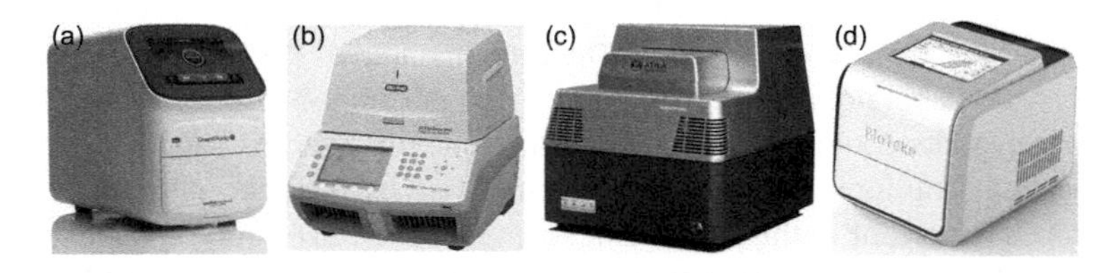

Figure 5.4 (a) QuantStudio™ 3 Real-Time PCR System, 96-well, 0.1 mL from Thermo Fisher Scientific, (b) Biorad Real Time PCR Machines, (c) Powergene 9600 Plus Real-Time PCR System, and (d) BioTeke Fast real-time fluorescence PCR analyzer (BTK-8).

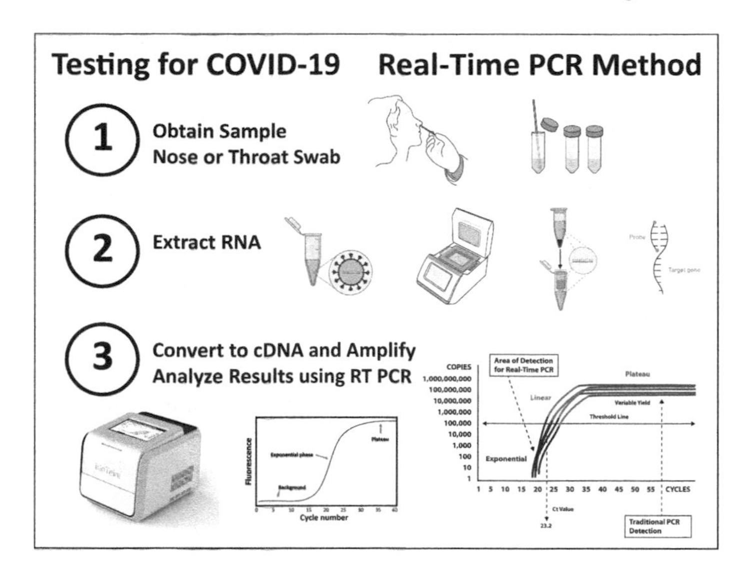

Figure 5.5 The COVID-19 test starts by taking a sample from the patient's nose or mouth. Samples are taken from these locations because they are easily accessible and are known to contain the virus.

fluorescent reaches an appropriate level to be detected the sample is positive, otherwise it is considered negative.

The data is then exported to a computer and processed. If the sample shows a level of COVID-19 RNA above the concentration (C_t) (negative sample threshold (cycle threshold)) and all of the controls produce the expected results, the patient tests positive for COVID-19.

5.3 Discussion of PCR methods

PCR techniques are a valuable means to identify a specific microorganism and for multiplexing many sequenced microorganisms. Frequently this information can be used to determine the phylogenetic relationships and, in this manner, organize all organisms according to their genetic relationships. PCR is used to amplify enough DNA or RNA for many other uses.

PCR is most useful when detecting known or sequenced microbes. Unknown microbes, those not sequenced, create a challenge. Some methods use primers that represent a higher classification primer, those that are unique for the order or family. In this manner, we have at least an idea of where in the phylogenetic tree an unknown microbe might be located, such as the *Paramyxoviridae* viruses or the hepatitis viruses. A truly unknown virus, and there are likely trillions, would be invisible to PCR until the unknown microbe is sequenced and a proper primer made.

Viruses that are manmade are a special situation. Identification can be confused, as a virus may be made to look like an existing sequenced microbe, when it actually has properties of something else. Until properly sequenced, primers made, and the proper phylogenic relationships determined, viruses may evade detection. Commercial PCR kits are available for quantitative analysis of a limited number of clinically important viruses. Some single-use PCR kits are being used for home testing.

Additional processing time is required before this test can be done when sequences, primers, and other procedures are needed, as in the case of a new virus. When these other steps are performed, careful skilled attention is required; this is not a fast process. The n+1 rule is important to remember; there is always one more virus.

5.4 Antibody methods

Frequently these methods are considered serology tests, that is, tests based on the presence of an antibody to a particular microbe. Usually, this is used to test for infections to which people or animals have been exposed. Often these antibodies will remain with a person for a long time, depending on the microbe.

5.4.1 Detecting viruses using antibodies and how do they work?

A blood sample is used to test for antibodies, although saliva or nasal secretions can be used. If there are antibodies present for a particular virus, they will match and bind to the viral antigen. The test determines if a patient has had previous exposure to a virus and has made antibodies. A negative result can result if taken early and the body has not had a chance to produce antibodies (Figure 5.6).

If antibodies have been produced for a particular virus, a marker (viral antigen) will react with them. The reaction results in a chemical change and a resulting visual signal. An example of this is the home pregnancy test; if there is a reaction to certain antibodies then the test turns positive. If there are no antibodies, or low values, there is no reaction. The body produces antibodies to an infection. Harvest these antibodies and put them on a probe and then expose them to the virus and you get a reaction, and a visual positive. The reaction depends on the antibody produced, for

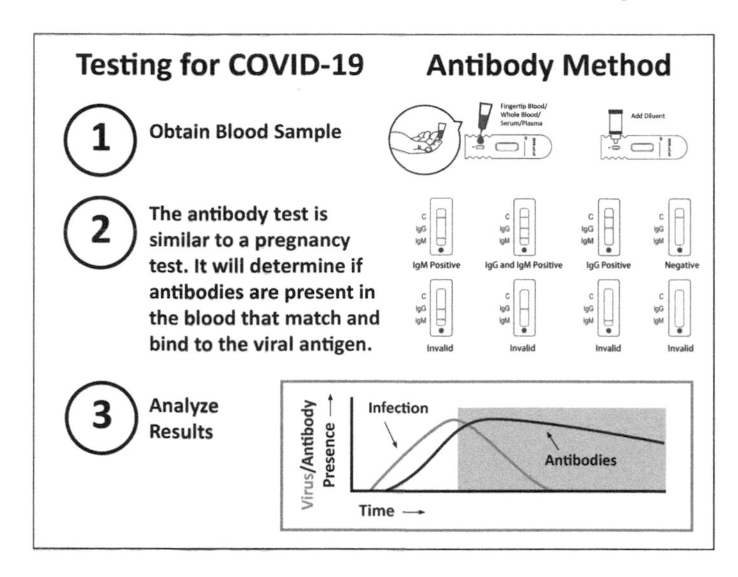

Figure 5.6 Antibody detection process for the COVID-19 virus.

example, if an antibody is produced for a portion of the virus, then the reaction will be for that portion, e.g., spike protein for COVID-19.

5.5 How do you add new viruses to the antibody method of detection?

Similar to other biomolecular methods, the first thing needed is to acquire the antibodies to a particular virus. Once these are acquired in enough numbers, they are incorporated into antibody test methods. Antibodies reacting to a virus antigen will produce a positive reaction. If there are no virus antigens, then there is no reaction and the test is negative. Adding new viruses takes time and skill.

6

Direct Virus Counting Methods, Such as IVDS

6.1 Introduction

Advances in the understanding the physical features of viruses and other sub-micron sized particles were made along with the advancements in particle counting technology during the late 1990s that allowed the counting of individual viruses. Viruses exist in a special physical space known as the Virus Window (Wick 2015), where the heaver particles are excluded and the lighter particles are also excluded; this was determined by their cesium density gradient determination which separated the viruses according to their size and according to their buoyant density. The different sizes were confirmed by electron microscopy. Furthermore, the concentration of these other particles, if present in a sample, are frequently low and below background measurements. The first instrument developed was the Integrated Virus Detection System (IVDS), including an electrospray, differential mobility analyzer (DMA), and condensation particle counter (CPC). This arrangement of instruments allows for the separation of the viruses and can "count" several viruses in a single sample. The process generally has simple sample preparation and takes just minutes to analyze. Viruses in a saliva sample can be analyzed in minutes.

Considering the many shapes of viruses from long thin ones, to small round ones, to polymorphic forms, the question is often asked, "How are the viruses seen on the particle counter?" The viruses, regardless of shape, are seen as an average size. The best comparisons are national standards made by precipitation. They have an average distribution which represents the standard, such as 50 nm. This size standard has a wide bell-shaped distribution with the center at the size standard, for example 50 nm. The same is the case for a virus; it is a distribution with the peak at a given size for that particular virus. It remains the same for the analysis. Viruses that are round present a narrow distribution in size and the bell-shaped curve is steep, such as the polio virus.

The IVDS method for detecting viruses changed everything with respect to virus analysis. There is no culturing and no reagents, the analysis is quick,

DOI: 10.1201/9781003106623-6

and it is generic to all viruses. It has a long shelf life, as the instrument can be stored on the shelf and plugged in when needed. It counts whole virus particles, is suitable for counting all virus types, and thus can also count unknown viruses or viruses we do not have other means to detect. The beta instrument was used to examine viruses separated from complex media with sensitivity that approached 10^4 viruses per milliliter; current instruments are improved (10^2). Other virus characteristics were observed and calibrated, for example the unexpected passage of viruses through filter membranes thought to be able to filter viruses (Wick 1999a) and the unexpected survival of MS2, a bacteriophage, in both extreme temperature and pH environments (Wick 1999b). It is the 4-nanometer separation of viruses that aids in this direct structural characterization of a large number of viruses. Some viruses have empty capsules and full capsules. These are frequently of different sizes, and this can be used to determine the ratio of empty and full capsules.

6.2 How does direct counting work?

The process is to prepare a sample following the outline in Figure 6.2. Clean samples can be introduced directly into the instrument and the viruses counted. Viruses are passed through an electrospray (ES), then a DMA, and then counted by a CPC. The electrospray puts the liquid sample into a condition to pass into the DMA which analyzes the sample and separates the various particles. These particles are then counted by a CPC, a device that coats nanometer-sized particles to form a micron-sized particle that can be counted by a laser. This process takes minutes.

6.3 How do you identify a new virus with direct counting?

Unlike other detection methods new viruses are simply counted by the IVDS instrument with no special processing. Unknown viruses are counted, their size and concentration determined. Other methods are used to phylogenetically classify the new virus. If it is not sequenced or truly unknown, this may take some time; meanwhile the virus has been detected and the size and concentration determined.

6.4 Why IVDS was invented

A new method for the detection and characterization of viruses had been needed for years. A method that detected the presence or absence of viruses was the primary requirement, and this need reached new emergency status as the rapidly expanding number of new and emerging viruses taxed previous detection methods. Adding the cost of previous methods to analyze large numbers of samples, a clear and urgent situation developed. A new means for detecting the presence or absence of viruses has been developed that capitalizes on the fundamental physical properties associated with these tiny microbes. The IVDS utilizes regular methods to purify and concentrate samples for sizing and counting using the methods of ES, DMA, and CPC. IVDS is essentially a virus particle counter that has been invented, patented, demonstrated, and placed into commercial use for a wide range of viruses both with envelope and without envelope.

This invention was possible because of two major features of viruses: they have different sizes for different families; and they have different physical properties from other submicron-sized particles in the environment. The first feature has exceptions, such as the polymorphs, but was found not to be a limitation because of their other features and scarcity in most environments. The second feature, the Virus Window, is exceptionally useful in that the virus families, particularly those pathogenic to man, can be separated by their physical features. These two features allowed the isolation and characterization of many viruses in the natural environment as demonstrated by the monitoring of viruses in honeybees.

IVDS improved on other methods in that it has: (1) no chemistry or biochemical requirements, (2) no shelf-life issues, and (3) the ability to detect all the viruses in a sample in a single pass without error. False negatives and false positives are unlikely.

One of the first tests of this new technology was to compare it with other methods, such as small angle neutron scattering (SANS). Results were comparable (Kuzmanovic et al. 2003). A second objective was to demonstrate the effectiveness of IVDS in counting viruses under a wide range of conditions. This is important as there are many inherent challenges to virus detection and analysis; among the primary ones is purification and concentration from the background material. This was accomplished for several viruses from many environmental samples, such as different soils, drinking water, seawater, and plants.

These early results were improved upon over the next ten years, and the second instrument, the first commercial IVDS, is still in use having proved to be a rugged and reliable device used to monitor honeybee viruses. The future instruments are proposed to be small, portable, automatic devices with greater resolution. It could be expected to see these in the local drug

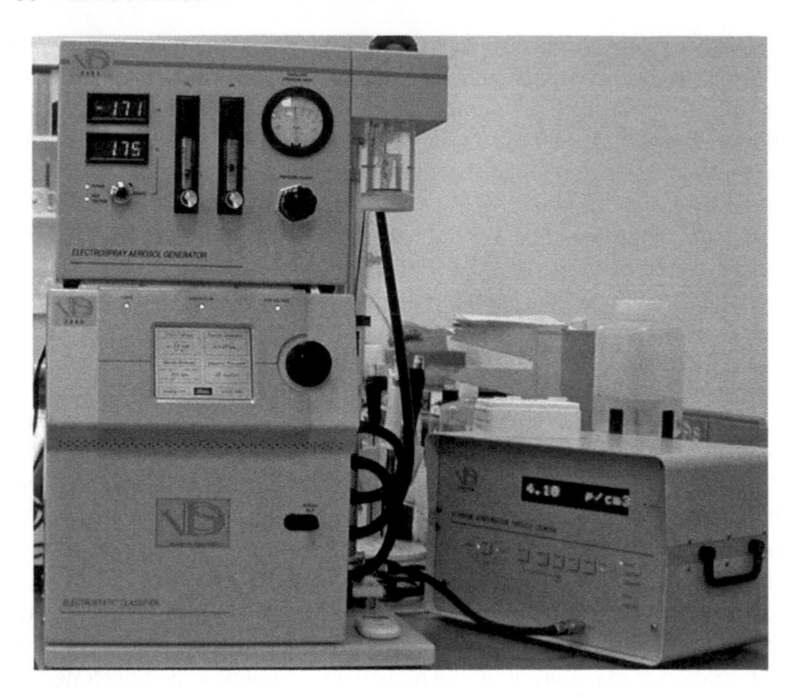

Figure 6.1 IVDS instrument.

store in use as a self-testing device. Figure 6.1 illustrates the IVDS instrument in use.

6.5 Flow chart showing how to use IVDS for virus detection

Figure 6.2 illustrates the various ways to use IVDS. This flow chart should look familiar to most biological sample processing with the exception of the detailed molecular biology steps, which are unnecessary for mechanical detection methods like IVDS. Each of these steps or stages has choices depending on the type of sample. In the simplest of steps this flow chart is reduced to three or four processes to yield detection and may only take a few minutes. More steps are required for complex samples.

The collecting stage has three choices; if the sample is a "clean" sample, it can proceed to the second stage, but if it is an air sample or a dirty liquid sample it needs processing. For example, for a virus that is already in a clean liquid, no further collection is necessary. Viruses need to be removed from air samples and transferred into a liquid. Since virus loading in open atmosphere is frequently low the sampling needs to be carried out over

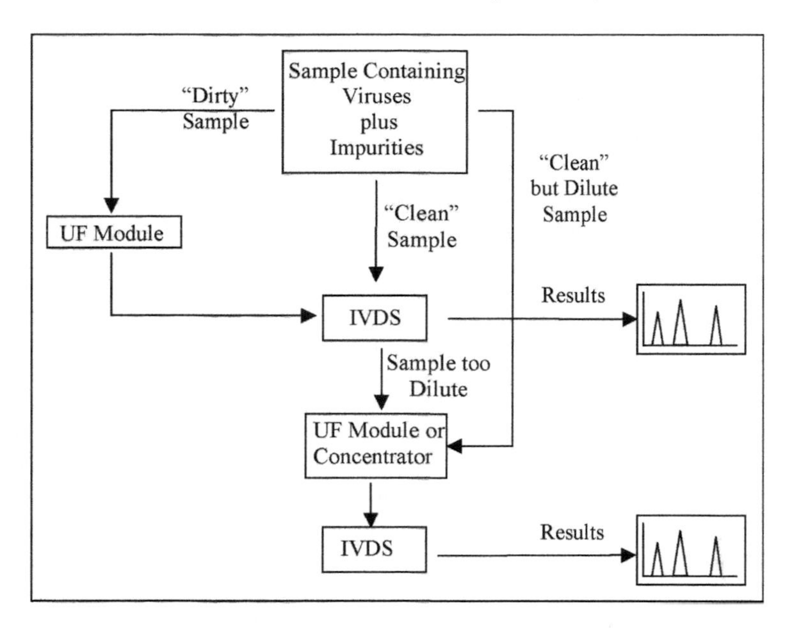

Figure 6.2 Flow diagram for sample processing in IVDS.

enough time to transfer the virus into a liquid. A liquid sample may need processing. A few viruses in a gallon of water are not necessarily evenly distributed among the gallon of liquid such as chemical compounds. The gallon samples need to be concentrated.

There are two main methods for purifying and concentrating viruses – filter methods and centrifugation. There are other methods available, but these are frequently much more difficult and time consuming and will not be discussed further. Multi-stage filtration can remove large background material and successfully isolate a virus and then further purify the virus by filtering the sub-virus sized material from the solution. This is the preferred step in this stage and in most applications works well with minimum effort and time. Centrifugation takes specialized equipment, and depending on the sample either a simple centrifuge or a more complex yet very useful ultracentrifuge can be used. This is the stage where the gallon of liquid is reduced to a few milliliters in volume, usually by size differential filtration. This processing of large volumes of liquid has been successfully completed for large water samples. The virus sample, after the completion of one of these steps, is reduced in volume, and the viruses are separated from the larger sized particles, concentrated to some degree, and ready for the next stage.

Remembering that the object is to count the viruses, it is understood that background material needs to be removed first. If the objective is to count the nuts in a jar of chunky style peanut butter it is frequently easier to remove the peanut butter first. This can easily be accomplished by placing the jar in hot water, waiting until the peanut butter melts, and then pouring the contents through a screen of a size to retain the peanuts. Then count the peanuts. The same analogy is applied to many different backgrounds.

A second purifying and concentration stage is indicated for samples that need to go through additional concentration and purification which can simplify the counting later. This step is frequently not needed in practice as the first purifying and concentration stage is usually sufficient.

During the detection phase the sample is ready to be inserted into the IVDS instrument. IVDS then examines the sample and reports any virus in the sample. This output will include all the virus and virus-like particles in the sample and will be sent to the computer for cataloging and further processing to yield size and concentration. This step is measured in minutes.

Sample information is taken directly to a computer. At this point, software can be used to redirect the results via the internet to multiple sites for use by decision makers, further analysis, or archiving.

The steps in using IVDS consist of the following:

- Collecting a sample (this is generally trivial in the case of waterborne particles, but can be nontrivial in the case of airborne particles or complex media, Figure 6.2)
- Purifying the samples by concentration and filtration
- Detecting the viruses

6.6 The recommended uses of IVDS

IVDS is intended to provide a system for the universal monitoring and sampling of viruses. Another objective of the IVDS invention was to provide a method and device for the rapid detection of viruses and their initial classification by size and concentration. This is based on the physical characteristics of viruses and therefore does not require the use of biochemical reagents or assays. It was a further objective to detect known and unknown or mutated viruses.

It was an important objective to be able to detect many viruses at the same time and thus provide a method to monitor the natural virus loading over time. This feature allows changes to be noticed in the virus loading and further gives information useful to the evaluation of the biological loading of viruses at any given time over a temporal period. Changes stand out from the natural or normal conditions and are immediately recognizable as a reason for alarm or alert and action.

Another aspect of IVDS is that the method further comprises counting the extracted and purified viruses and classifying according to size and concentration. This is important because viruses tend to decompose during an infection.

Although to a very large degree only viruses fall within the Virus Window, other background components fall close to the Virus Window. These components are microsomes and similar sub-cellular structures. These components can be effectively eliminated by adding nonionic surfactant, such as diethylene glycol monohexyl ether, to the collection stage.

IVDS is a first line detector to determine the presence or absence of a virus. We have many other methods that can be used later to confirm or identify a virus, and during this second step the process then provides both virus detection and confirmation by two different technologies. IVDS can be used to screen large number of samples since it is quick and can detect multiple viruses in the same sample.

Initially and as described below the full invention of IVDS was more robust. When the ultracentrifugation step is included in the integrated system, it is possible to identify the virus later detected by the DMA unit because of the unique 3-D address given due to the physical properties of the viruses. In practice, this step is frequently not used. The result is that IVDS can be used to quickly detect the presence of viruses in a sample and estimate their concentration. This information has been most useful and in practice all that is generally needed to make management decisions.

The presence or absence of a virus is the first question of a first responder or someone making an analysis of a biological sample of interest. The detection of a particular sized virus then is useful in making a decision to test further. In the case of an infectious particle further decisions can be made in regard to isolation and quarantine or other management decisions while further tests are conducted, tests that frequently take much longer, the results of which could take hours or days. Frequently, we do not have the time to wait for results to make a decision, as an infectious agent can spread rapidly. Also, we do not have the luxury of isolating or quarantining everyone who may be infected or of interest.

The IVDS instrument relates to the detection, size determination and concentration, and monitoring of submicron size particles, but mainly viruses. The other particles include such particles as prions, viral subunits, viral cores of dilapidated viruses, etc., in bioaerosols and fluids, especially biological fluids. These "other" particles are generally not an issue in detecting intact viruses.

The difficulty of detecting and monitoring a wide range of viruses also varies by environment, and perhaps the most troublesome environment involves combat conditions. In particular, the problem of detecting and monitoring viruses in a potential biological warfare (BW) threat environment is extremely demanding. Variations in virulence from virus to virus are generally accepted, and ingestion of 10^4 virions constitutes a significant threat to a soldier who breathes on the order of 1,000 liters (1 m^3) of air per hour. Instruments with sensitivities which enable the detection of remote releases of biological agents in a field environment, thereby providing early warning capabilities, allow important safety and operational decisions to be made. Although progress has been made on many fronts, such as optical means and other such approaches, such an instrument remains elusive.

In the past it has been difficult to maintain broad-spectrum systems for the detection of viruses which are free from false negatives because of natural or artificial mutations. The high mutation rates of known viruses, as well as the emergence of new viruses, such as the Ebola virus, must be addressed by a detection method. There is also the potential for deliberate artificial mutations of viruses. There are virus-like infectious agents, such as prions, which are suspected in causing scrapie, "mad-cow disease", and Creutzfeldt-Jakob disease. These prions possess no DNA or RNA, and can withstand 8 MRads of ionizing radiation before losing infectiousness. Other virus-like infectious agents, such as satellites, possess no proteins. However, detection of all of these agents must be possible for a device or method to be generally effective in the detection and monitoring of viruses and similar agents (such as virus fragments, prions, satellites, etc.) which are pathogenic.

The detection and monitoring of viruses must also be free from false positives associated with various and diverse backgrounds. The background includes biological debris which obscures the detection of the viruses by registering as a virus with the detection methods used in analyzing the samples collected.

The analysis of viruses requires a very high degree of purification of those viruses to overcome background loading in order to avoid false positives. For example, a BW virus may be buried within loadings of other microorganisms which form biological debris with loading on a magnitude of 10^{10} larger than the threshold loading for the targeted virus itself.

Although methods that culture viruses can often be used to increase the virus over background, culture methods are too slow for efficient viral BW detection. Some important viruses cannot be cultured by known methods, and in any case cell culture is a highly variable and inconsistent method.

Viruses may also be extracted from an environment and concentrated to an amount that is required for detection and monitoring, without requiring any culturing. For the detection of small amounts of viruses in environmental or biological liquids, it is necessary to both enrich the concentration of viruses by many orders of magnitude (i.e., greatly reduce the volume of liquid solubilizing the viruses) and accomplish the removal of non-viral impurities.

Sampling for airborne viruses is generally accomplished by collecting airborne particles into liquid, using a process such as air scrubbing, or eluting from filter paper collectors into liquid. Since collection and subsequent separation and detection methods are strongly affected by the adsorption of viruses to solids in aerosols and by solids-association in water, this poses stringent requirements on the design of the sampling of air for viruses.

In contrast, when sampling liquids for viruses, no special equipment or processes are generally necessary in order to collect a sample; for example, in sampling blood for viruses, only a standard clinical hypodermic needle may be needed, and similarly for other body liquids. For the sampling of bodies of water or other conveniently accessible liquids, sample collection may not be an issue at all, and in such cases the term "collector" is often applied to what is, in reality, a virus extraction step (such as collection on a filter).

Currently, only IVDS is available for the detection of viruses in a BW environment. Rapid detection translates into protection and more reliable and simplified strategic planning, and the validation of other BW countermeasures. Previously known detection methods using biochemical reagents are impractical in the field, even for trained virologists. Reagent-intensive approaches, such as multiplex PCR, low-stringency nucleic acid hybridization, and polyclonal antibodies, may increase the incidence of false positives by several hundred-fold. The hypervariability or rapid mutation of viruses and emergence of new, uncatalogued viruses further preclude methods based on biochemical assays, such as PCR, immunoassay, and the like, from achieving broad-spectrum detection of all viruses regardless of identity, known or unknown, sequenced or unsequenced. There is a need for a highly reliable automated system.

6.7 Improving the sensitivity of IVDS (concentration and accumulation)

A concentrator device and method of concentrating a liquid sample was invented (Wick 2012a). This apparatus, illustrated in Figure 6.3, replaces the sample chamber on the IVDS. The concentrator device functions as a pressure vessel that contains a slant tube with a filter element. Air, or gas, is introduced into the pressure vessel and this pushes the sample through the slant filter, effectively concentrating a 1 milliliter sample to a few drops, less than 1 microliter for a 1000:1 concentration. The sample is then introduced into the ES in the regular manner.

Figures 6.4 and 6.5 illustrate the effectiveness of the concentrator, and this makes for a more sensitive instrument. A small 1 ml sample can be concentrated to 0.1 ml.

Likewise, the sample can be accumulated over time. The usual counting time for a sample is one minute. It is possible to count the sample several times and sum the results to increase the virus count. Figure 6.5 illustrates the results of counting a MS2 bacteriophage for 1–10 minutes.

In this manner the sensitivity of the instrument was improved, providing a 1000:1 increase by concentration and an addition 10:1 by counting longer. Counts of more than 100 minutes have been demonstrated, in this manner increasing the sensitivity by 10^5 (Figure 6.6).

6.8 An example – following COVID-19 through 5 days and then a 3-month follow-up

The sample was collected from an individual over 5 days and again after 3 months. A sample of saliva was collected in a small bottle each day; it was filtered through a 0.45 μm circular syringe filter to separate large pieces, then diluted 1:1 in reverse osmosis (RO) water and introduced into the IVDS for analysis. Sample time took about 4 minutes. In this case, two 1 minute 45 second scans were made. Recall the COVID-19 virus as seen in Chapter 3: an illustration of the virus and the attached spike protein as seen by electron microscopy. There are three charts: the first illustrates the virus core at 73.5 nm, the second illustrates the spike protein at 31 nm, and a third chart

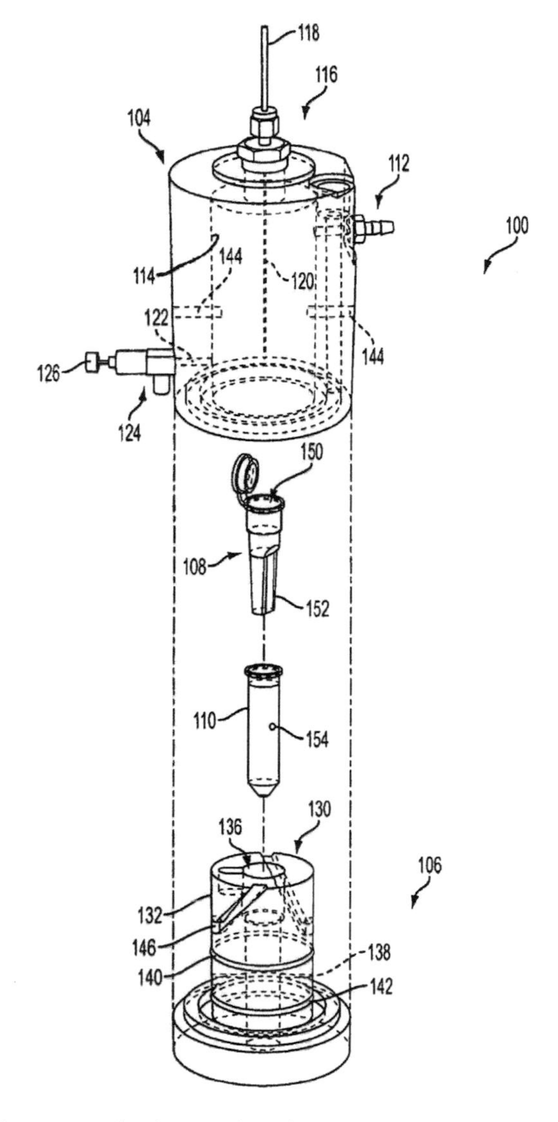

Figure 6.3 Concentrator, showing a 1 ml sample tube. A sample can be concentrated from 1 ml to 0.1 ml.

that illustrates the breakdown of the spike protein (just the tips) at 14 nm, a size that is typical for saliva samples.

Day 1. This day was pre-symptomatic. The COVID-19 virus core is shown at 73.5 nm (Figure 6.7); it manifests without the spike protein which makes it appear larger. Twelve counts are recorded. The spike protein is illustrated

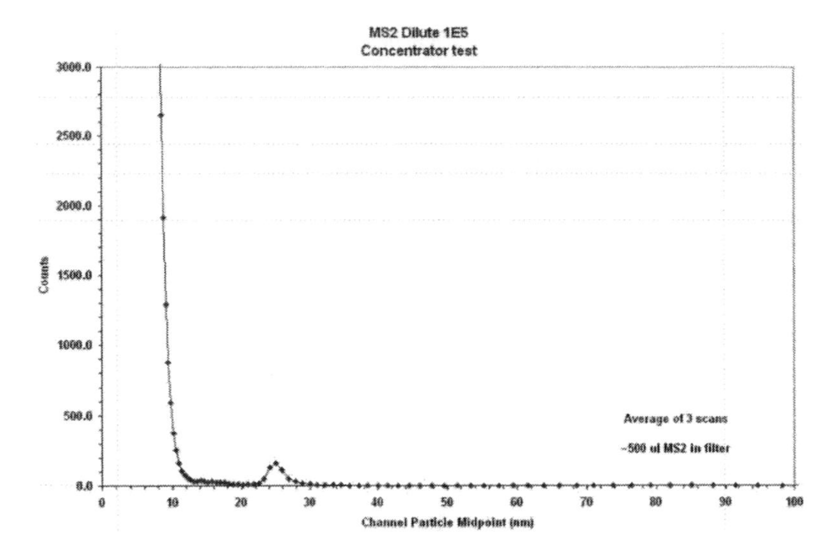

Figure 6.4 MS2 sample prior to concentration.

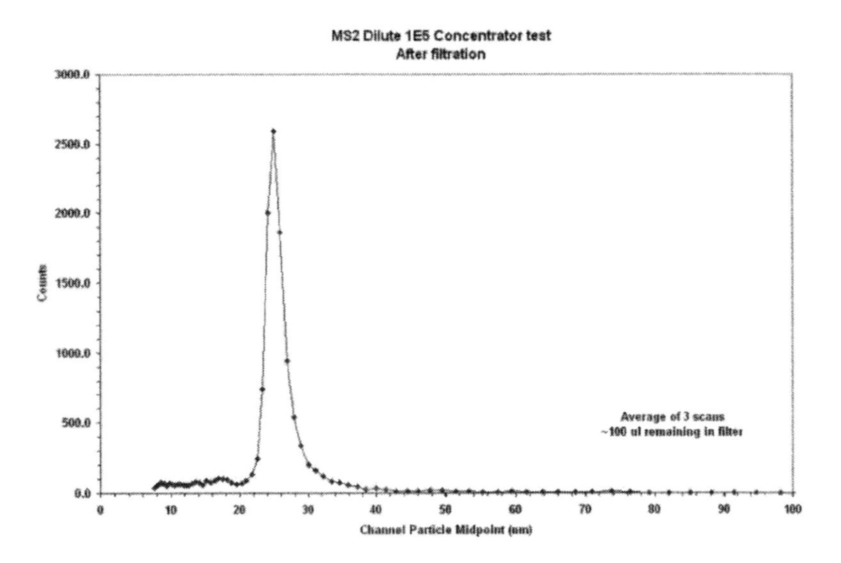

Figure 6.5 MS2 sample after concentration.

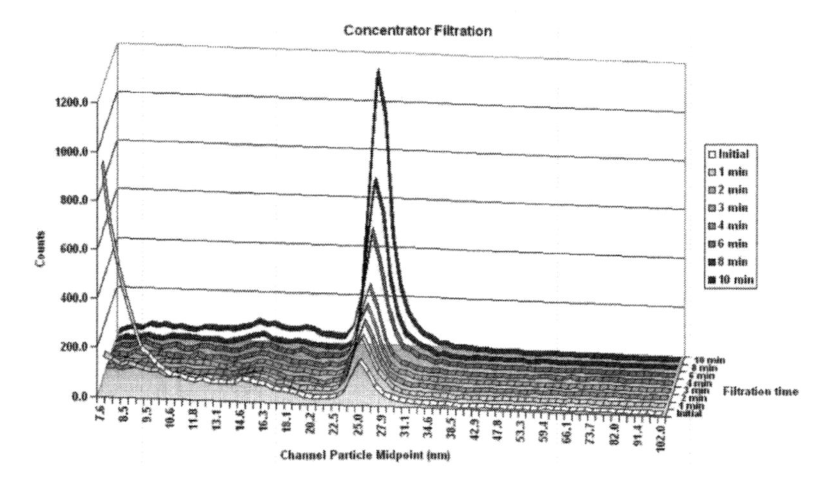

Figure 6.6 Virus sample after counting from 1 to 10 minutes.

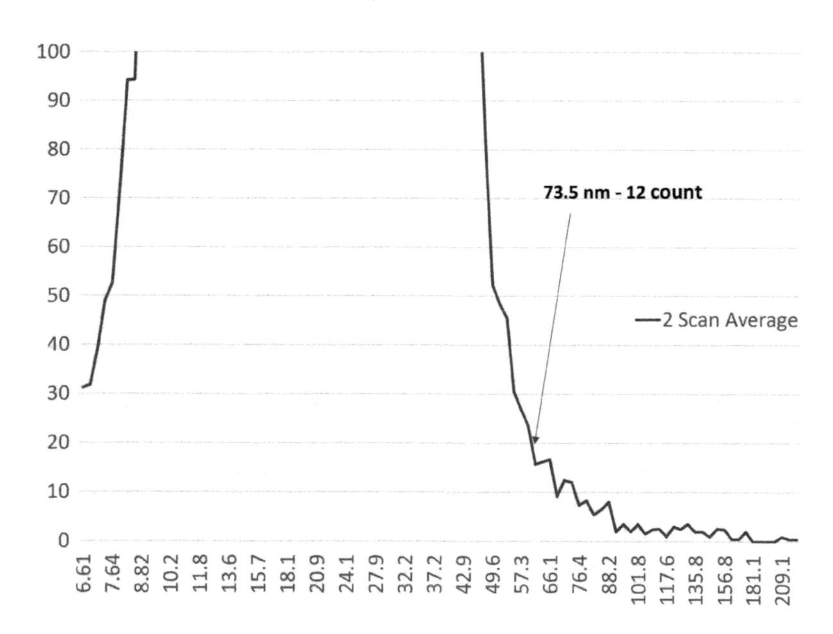

Figure 6.7 Day 1 of a COVID-19 infection showing a count of 12 for 73.5 nm.

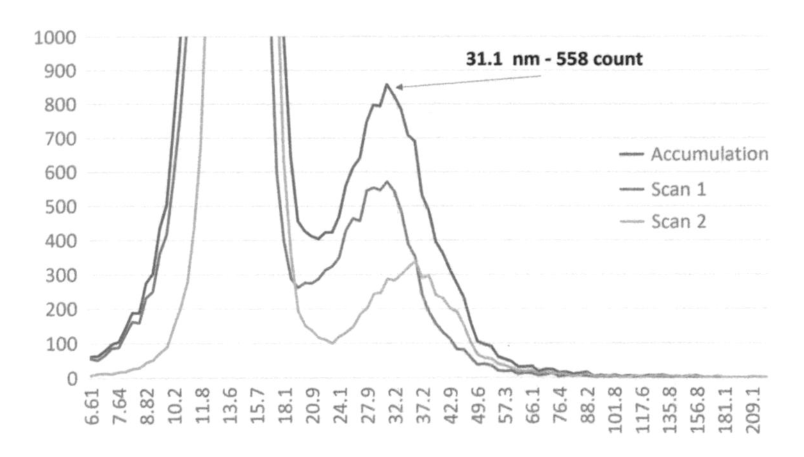

Figure 6.8 Day 1: the detection of the spike protein, showing the typical appearance of an aggregation of spike protein components.

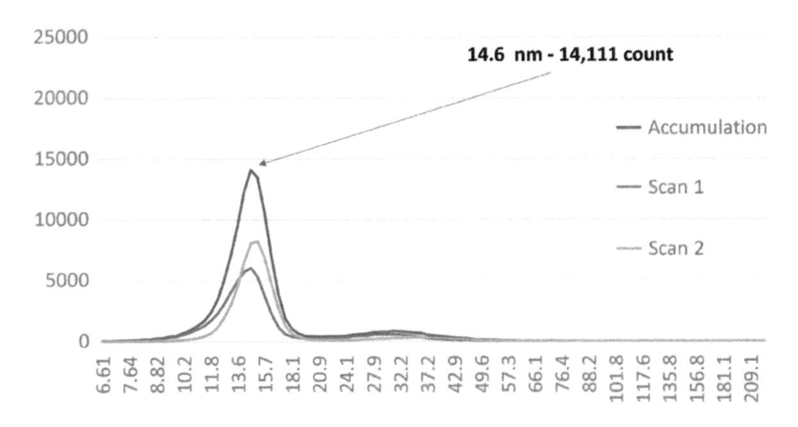

Figure 6.9 Day 1 showing the breakdown of the spike protein; it is a typical detection size range of saliva.

at 31.5 nm with a count of 558 (Figure 6.8). The spike protein is seen disassociating into 14.1 nm particles with a count of 14,111 (Figure 6.9). This size is also seen in a typical saliva sample, and further analysis may be needed to verify the spike protein.

Day 2. This day was symptomatic, with discomfort, the start of a common cough and lack of smell and taste. The COVID-19 virus core is shown at 73.5 nm with 18 counts (Figure 6.10).

The spike protein is illustrated at 39.5 nm with a count of 732 (Figure 6.11), which is an increase from the previous day; it is seen as a wide peak because it appears to be starting to dissociate.

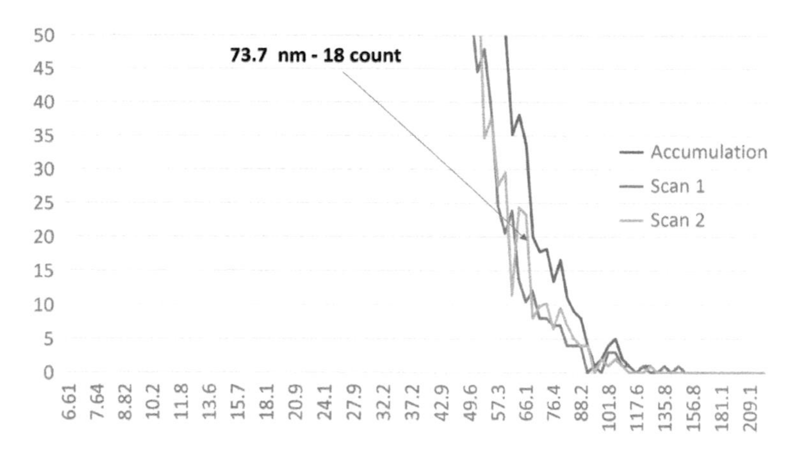

Figure 6.10 Day 2 showing 18 counts for the 73.7 nm COVID-19 virus.

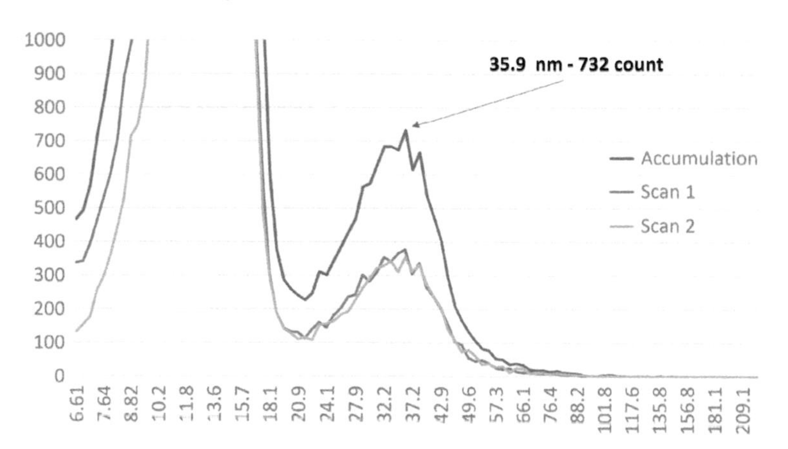

Figure 6.11 Day 2 showing the 732 counts for the 35.9 nm COVID-19 spike protein.

A typical detection size range of saliva at the 14.1 nm peak is seen in Figure 6.12.

Day 3. This day was symptomatic, with strong discomfort and fatigue, a stronger cough, and lack of smell and taste. The COVID-19 virus core is shown at 71.0 nm with 189 counts which is a large increase from the previous day (Figure 6.13).

The spike protein is illustrated at 28.9 nm with a count of 4,884, indicating the continued disassociation of the spike protein to the predicted final size of 20.9 nm (Figure 6.14).

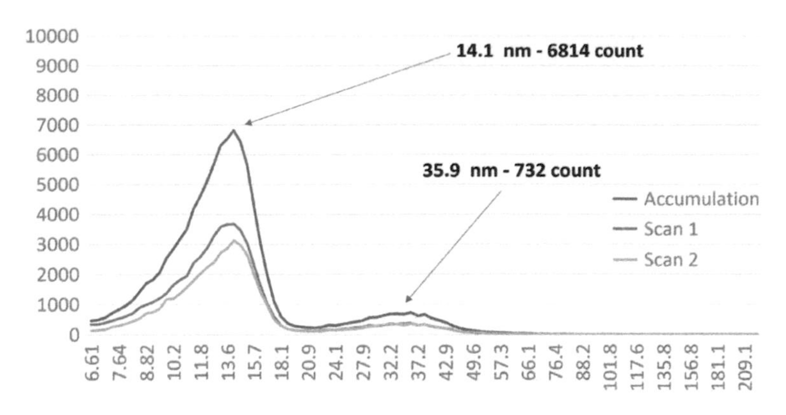

Figure 6.12 Day 2 showing the saliva and the continued dissociation of COVID-19 spike protein.

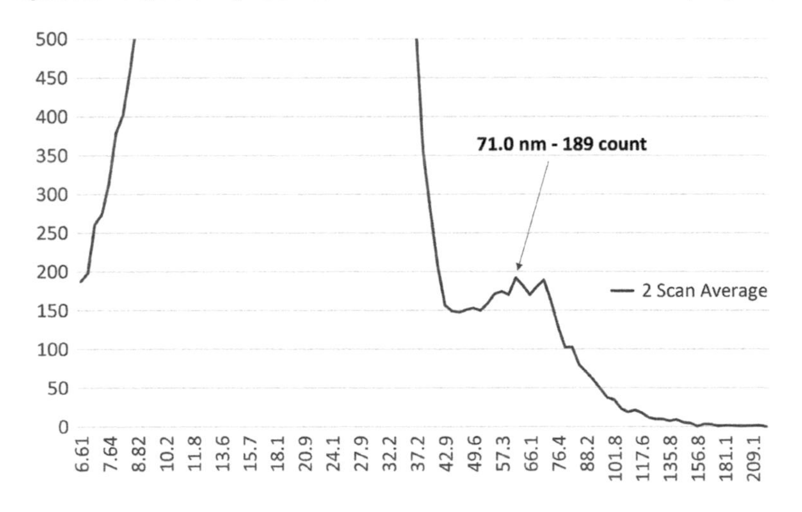

Figure 6.13 Day 3 showing the 189 counts of the 71 nm COVID-19 virus.

Day 4. This day continues symptomatic. The COVID-19 virus core is shown at 73.0 nm with 0 counts (Figure 6.15).

The spike protein is illustrated as a wide peak at 22.3 nm with 28,853 counts (Figure 6.16).

Day 5. This day was asymptomatic, other than continued fatigue. The COVID-19 virus core is shown at 73.7 nm with 1 count (Figure 6.17).

The spike protein is illustrated at 23.3 nm with a count of 17,803 in Figure 6.18. The scans represent the two 1-minute examinations by the DMA. It

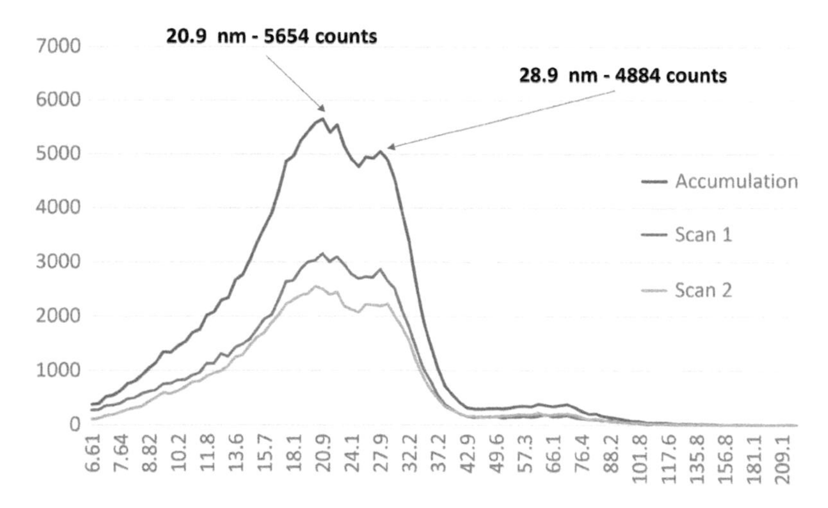

Figure 6.14 Day 3 showing the spike protein.

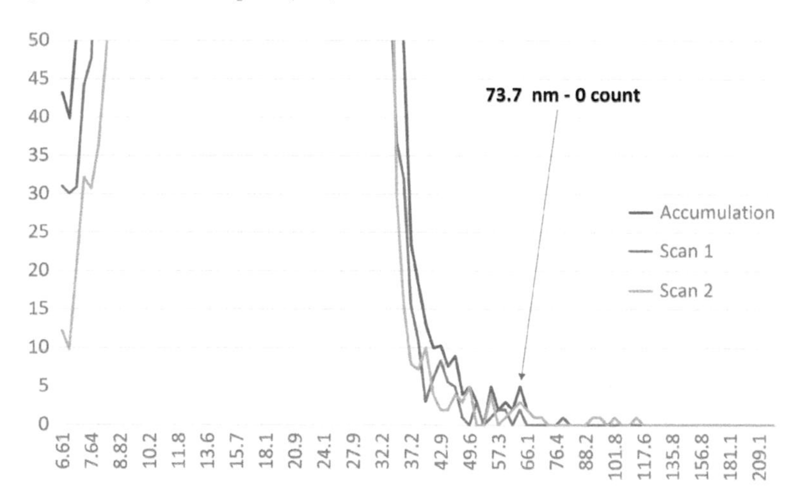

Figure 6.15 Day 4 showing the absence of the COVID-19 virus.

is thought that the higher counts are the continued disassociation of the virus.

Samples were taken 3 months later, following a day absent of symptoms. A peak is shown at 55.2 nm with 542 counts. This would still indicate the spike protein aggregating and disassociating to the 20.9 nm illustrated in Figure 6.19.

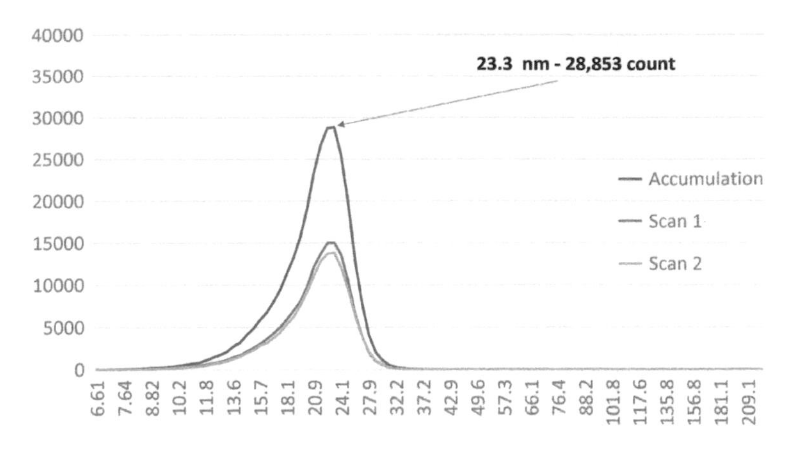

Figure 6.16 Day 4 showing the spike protein.

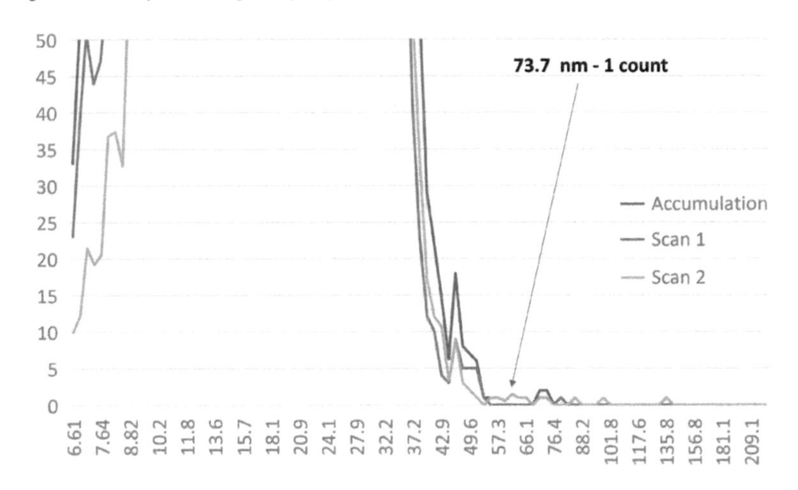

Figure 6.17 Day 5 showing the one count of the COVID virus.

The spike protein is illustrated in Figure 6.20 at 20.9 nm with a count of 74,543 after 3 months post-infection. This type of information may be useful for antibody probes designed for the spike protein.

This observation of COVID-19 over a period of time is a typical example of the screening capability of IVDS. The COVID-19 virus core was detected early, pre-symptomatic, and it was possible to detect the virus during symptoms and still detect it after 3 months. Of particular interest is observing the break-up of the virus during this period. The increase of the spike protein at the onset is also of interest.

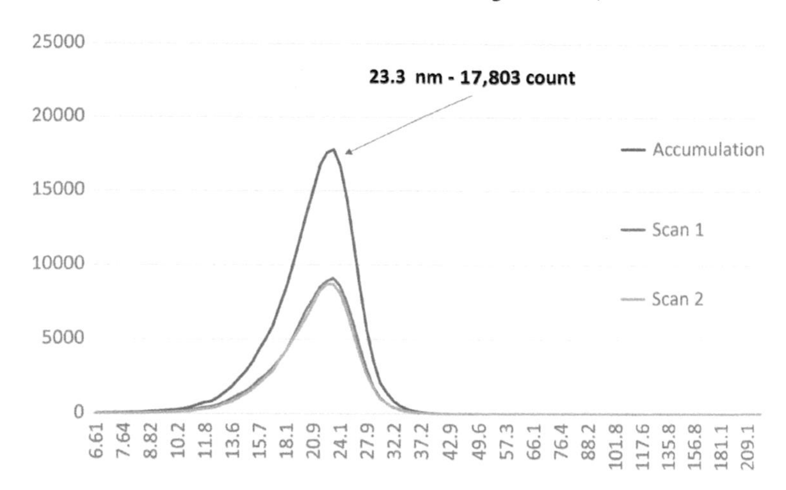

Figure 6.18 Day 5 showing the continuation of the COVID-19 spike protein.

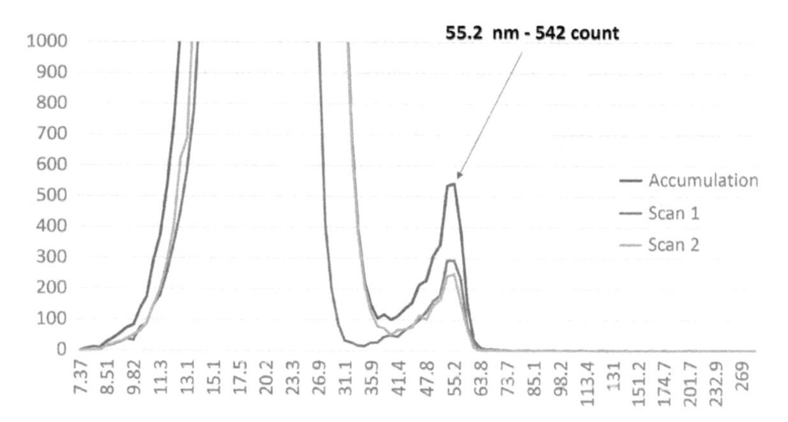

Figure 6.19 Particle at 55.2 nm after 3 months following infection with COVID-19.

The size of the intact COVID-19 virus is 128 nm; it is interesting that upon seeing an infection this quickly dissociates into the 71–75 nm core virus depending on how many of the spike proteins remain attached. Evidently the spike proteins come off early during an infection and continue to dissociate. These spike proteins come off the virus, and some of them group into aggregates, and then the aggregates continue to break apart into the smaller size. Both the aggregate and the individual spike proteins are seen. The spike proteins remain for a period of time after the virus is no longer detected.

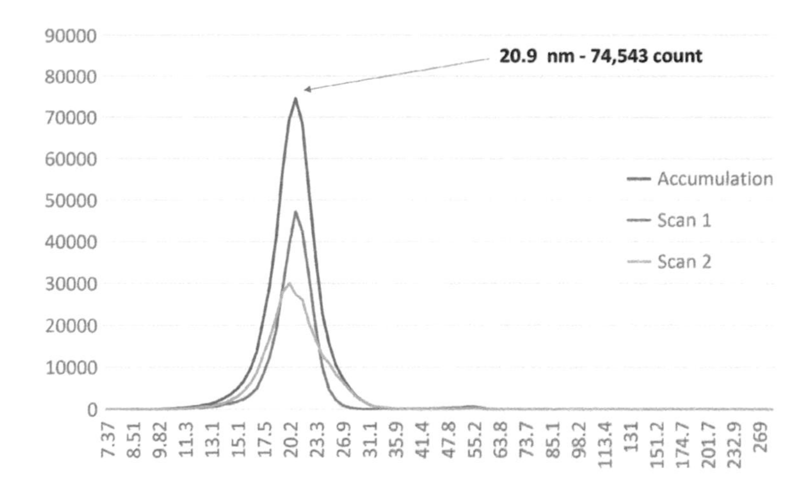

Figure 6.20 Three-month follow-up of COVID-19 infection showing spike protein.

The lingering of the spike protein is interesting. This sort of information has been seen for other viruses and continues to be studied.

6.9 PCR and IVDS compared

PCR is a great means to verify a particular virus if you have: (1) enough protein, (2) a target organism, (3) no contaminations in the sample that can interfere with results, and (4) negative results are difficult to confirm. When PCR detects an organism, the follow-on work of placing the organism in the phylogenetic tree and further classification work can then proceed. PCR probably should not be used for screening for viruses.

IVDS can detect viruses. The comments generally associated with the use of IVDS include: (1) does not identify the virus genetically. It does determine the size of the particle and gives a concentration. Depending on where the sample is collected the virus can be inferred. (2) Rapid sample preparation. (3) No reagents needed. (4) Is great for the screening of viruses because it does not target a particular virus. Multiple viruses can be detected in the same sample.

Both PCR and IVDS have their place when it comes to detecting viruses. One issue is the detection limits. PCR requires enough protein to get a good reading. Below this level is usually considered a negative result. IVDS appears to detect viruses below this limit (Figure 6.21). The 31.1 nm and the 38.5 nm particle are clearly detected in the honeybee sample shown in Figure 6.21. The same sample was negative for PCR.

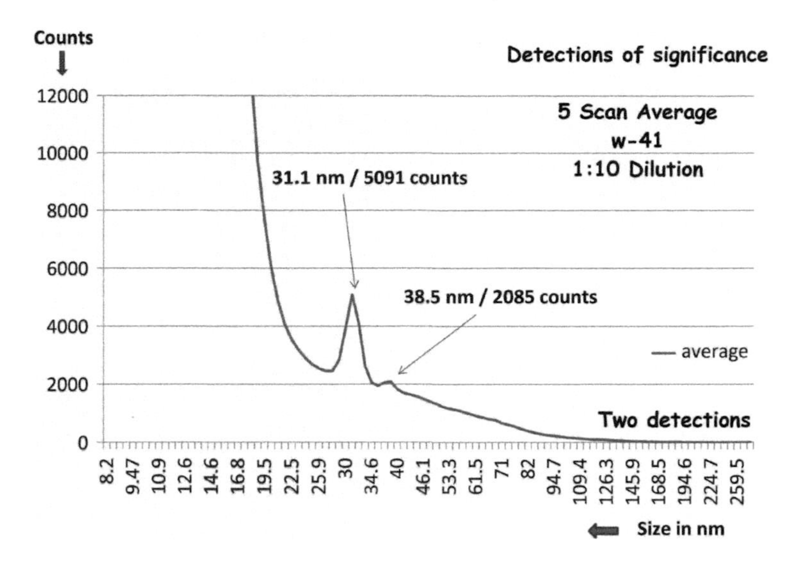

Figure 6.21 Detection of viruses not detected by PCR.

The accuracy of the PCR primers is a concern. When tested with enough protein, three different laboratories gave three different identifications for the virus particles. That is a concern when dealing with certain pathogenic viruses. IVDS detects the virus; the fact that there is a virus-sized particle present is concern enough, other methods such as PCR or mass spectrometry can sort out the details.

6.10 Summary of the fielded IVDS

- Detects intact viruses from 10 nm to 500 nm
- Detects viruses that fragment into components
- Is a stand-alone, portable device (usually made up of three components)
- Does not use reagents
- Can detect a virus in 5–10 minutes
- Requires minimum training to use
- Can detect unknown viruses (unsequenced)
- Can detect multiple viruses in single sample (see seawater example)
- Sensitive, demonstrated counts of 10^2–10^3 viruses (probably can be lower, but this is unnecessary in operation)
- Centrifuge not needed in operational conditions

7
Mass Spectrometry Proteomic (MSP) Method

Mass spectrometers have been used for the detection and classification of viruses. Many advances have been made in the hardware and software associated with mass spectrometers. Mass spectrometers are smaller, lighter, and easier to use. Computers have improved resulting in faster processing, and they are used in many disciplines. Although they were primarily used by chemists, biologists and other disciplines soon began to find uses for them. There are many types of mass spectrometers and they all have their uses; some of these are discussed. One type, an electrospray mass spectrometer-mass spectrometer (ES-MS-MS), has been used to detect and classify microbes, including viruses. It works similar to biochemical methods, in which biological sequences are downloaded from the National Database. Here methods diverge as the sequence information is sorted and all the unique peptides are determined for each organism. Sometimes these can number into the tens of thousands for each organism. The mass spectrometer then detects and identifies the unique peptides, and software determines the names and details of the microbes. Viruses are among these microbes. Once detected and classified the viruses detected can be archived for future reference in a computer file. The mass spectrometer file can be sent electronically to aid in analysis and decision making regarding the occurrence of detected viruses.

7.1 Introduction

Since the 1970s analysis by mass spectrometry methods has exploded into a wide and extensive network of scientific disciplines. The early vanguards were chemists and then bio-chemists and then biologists and then molecular biologists, and now it is difficult to say who is in the lead. Mass spectrometers are common in a modern laboratory, and nearly all academic areas have their mass spectrometers. It is now common to see a mass spectrometer center in major universities that services a wide range of academic disciplines. This has proved to be a viable solution as the state-of-the-art instruments continue to advance rapidly, and it is not unusual to see

DOI: 10.1201/9781003106623-7

substantial improvements in capability in months rather than years. A new graduate student can expect to start mass spectra analysis on one machine and finish on a more sensitive, faster, and more capable mass spectrometer.

Generations of mass spectrometers in a single laboratory have created follow-on applications of this technology in other areas, and as the operation and maintenance of these instruments continue to be simplified these new capabilities have possibilities. Advancements in software and computer speed have enabled faster analysis of mass spectrometry files, and often these files contain sufficient information for the classification of microbes.

The first step is to examine the different types of mass spectrometers. Scientists and engineers have been busy, and the result over the last 40 years is a variety of approaches to analysis by mass spectrometry. This chapter examines the major types of MS instruments and their various combinations used in microbe analysis.

Methods for sample preparation, instrument operation, analysis, and reporting of results are as varied as the types of platforms. It is appropriate to review those methods associated with the different mass spectrometers.

Because of this variability in the applications of MS, it is important to include the sample preparation methods which vary according to MS techniques. It needs pointing out that while it is possible to prepare a sample in under 10 min for inserting into the instrument, sample preparation time may be reflected in the dynamic range of the instrument. Fast sample preparation does not always indicate the best results. Recent improvements in dynamic range, however, by several orders of magnitude indicate that additional increases in the dynamic range can be expected for MS methods thus making short sample preparation times more attractive. Although not always indicated, it is assumed that sample preparation methods can be automated.

Instrument operation is a function of existing hardware, and possible future improvements can be expected. It takes a skilled operator to use the MS. Likewise, it takes a skilled person to prepare the sample for inserting into the MS. The current descriptions are generic in nature and due to the many types of MS systems, the demonstration of some small sized units and the development of mini-MS systems are ongoing concerns. This would indicate that the size and weight of future MS systems will be suitable for mobile operations. Hardening and manufacturing to military specifications should likewise be a straightforward process. The trade-offs between results, weight, power, and other requirements can be evaluated to create a future MS system that is optimized for specific applications, e.g., medical, emergency response, home use, and environmental applications.

Although mass spectral analysis is the process that has the greatest possibility for rapid improvement, it is the bioinformatic tools that have made it possible to quickly process large MS files in a short time. Such software improvements over the last few years have reduced processing time from hours to seconds. Similar improvements in hardware and software can be expected, and the additional speed will further reduce mass spectral data processing time.

MS techniques determine the molecular mass of compounds by separating ions according to their mass-to-charge (m/z) ratio; therefore, any species to be analyzed by MS has to be ionized in the first place. The ionization methods frequently used for the analysis of microbial constituents are described. Ions formed in the ion source of the mass spectrometer are next moved to the analyzer section that uses combinations of electric and magnetic fields to separate and detect ions according to their m/z values. Molecular ions formed in the ion source may be additionally excited and forced to dissociate into products which are then immediately mass analyzed. The most popular mass analyzers and their hybrids are used for microbial detection and identification purposes.

The following types of mass spectrometer are thus presented to simply demonstrate the various approaches to mass spectrometric analysis and the creative approach that scientists and engineers have taken over a short time.

Ionization methods used in mass spectrometry are suited for microbial detection and characterization. There are many types of mass analyzers. Remember that mass spectrometers were first used by chemists and that many of the approaches were from this line of thinking. In any case all the mass spectrometer approaches start with ionization. The evolution of the many types of mass analyzers and their different attributes is important as it provides choices in the types of mass spectrometers and how they work.

7.2 Ion mobility and various types of mass analyzers

Microbial agents are usually collected and undergo processing designed to obtain cellular constituents in a form suitable for ionization. The ionized species are mass analyzed and the data interpreted using computer-assisted methods.

The application of MS to any class of species is challenging because gas-phase ions are required for analysis. Ionization processes, which are used for the analysis of microbial constituents, are discussed in "Identifying Microbes by Mass Spectrometry Proteomics" (Wick 2013), and listed here, but will not be discussed further:

- Electron impact (EI) ionization
- Chemical ionization (CI)
- Atmospheric-pressure chemical ionization (APCI)
- Metastable atom bombardment (MAB)
- Electrospray ionization (ESI)
- Nano ESI-source
- Matrix-assisted laser desorption/ionization (MALDI)
- Surface-enhanced laser desorption/ionization (SELDI)
- Desorption electrospray ionization
- Electrospray-assisted laser desorption/ionization (ELDI)

7.2.1 Using the electrospray ionization (ESI) method in detecting viruses

General steps to prepare microbial samples for analysis by MS:

- Sample prep (for example, trypsin digestion for peptides)
- Ionization of analytes
- Analysis by MS
- Data collection and analysis

ESI has inherent analytical advantages that allow it to be utilized for the analysis of different biological problems. Since the limitation of molecular mass is minimal, relatively large biomolecules have been successfully mass analyzed using ESI techniques that include even intact viruses or their chromosomes (e.g., coliphage T4 DNA with nominal molecular mass of 1.1×10^8 Da) (Chen et al. 1995; Smith et al. 1994). Moreover, microbial carbohydrates, lipids, single stranded DNA, RNA, proteins, and peptides were studied through ESI-MS and have been used for the detection and identification and classification of microbes (Wick 2014).

A comparison of instrumental capabilities for various mass analyzers indicate that each has their favorable attributes and likewise limitations. These differences are suitable and optimized for different and various applications. It is important in the context of microbe detection and identification to consider the level of detection and identification; an application to simply

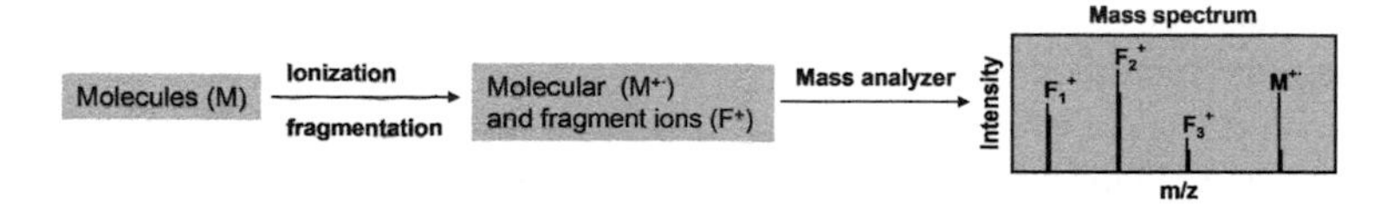

Figure 7.1 Measurement of mass-to-charge ratio (m/z) of molecular and fragment ions based on detection of mass analyzed ions by mass spectrometer.

detect a microbe is different than an application to identify a microbe to the strain level (Figure 7.1).

7.3 How do MSP methods work for biological detection?

The concept of operation is to collect a sample, prepare it for the mass spectrometer, have the mass spectrometer create a RAW (unprocessed data, it is a file extension) file and analyze the RAW file to determine the peptides detected, and use these peptides to determine which ones are unique to the various microbes. Thus, each microbe is detected by determining a match from the environmental sample peptides with those unique peptides, determined by calculation, based on the sequence of the microbe. In this manner all the sequenced microbes are associated with their unique peptides and matched with the sample peptides collected from the environment.

The following is an example of the physical process for testing samples of honeybees which were collected from various sources and processed for the ES-MS-MS:

- Step one, make bee smoothie
- Step two, prepare sample
- Recipe, trypsin digestion
- Run though MS
- Process RAW file using ABOid (software)
- Make report

Agents of Biological Origin Identification (ABOid) is an analytical tool invented and patented by the US Army (United States of America Patent No. 8,224,581 B1, 2012) (United States of America Patent No. 8,412,464 BI, 2013) and under license to BIOid, Inc. ABOid is fully described in Chapter 6 of a book by Wick (Wick C. H., *Identifying Microbes by Mass Spectrometry Proteomics* 2014). ABOid identifies a microbe by searching sample ion spectra of peptide ions against theoretical peptides determined by calculation from microbe DNA/RNA sequences. ABOid simply determines those unique peptides for each microbe and compares the unique peptides with those peptides identified in a sample. The result is a highly accurate, gene-based, identification of viruses, bacteria, and fungi among other sequenced organisms downloaded from the National Library of Medicine, National Center for Biotechnology Information (NCBI). ABOid is thus the tool of choice when seeking to identify large numbers of microbes in a sample. In the case of COVID-19 all the sequences of the various strains were downloaded from the NCBI and added to a data group and used to identify COVID-19 in a collected sample.

ABOid has been used to analyze thousands of environmental samples and has identified nearly a thousand microbes. This is particularly important when the challenge is considered: at this writing, there are about 254,000 Prokaryotes (bacteria) and 45,000 viruses and 12,100 Eukaryotes (fungi, plants) that have been sequenced. There are also 16,500 organelles and 22,000 plasmids. To make a new primer or spectra or antibody or similar need simply takes too long and costs too much money. These historic methods struggle to keep up, but can be used later when there is time to develop the special needs and control the costs.

Given the success of ABOid in identifying and classifying microbes in complex environmental situations and the ease of sampling, ABOid is a compelling tool to use anywhere there is a need for microbe identification that includes bacteria, viruses, and fungi. Since this can be accomplished in one sample, one effort, and at low cost with an archival record, ABOid has been selected as the analytical tool of choice.

It needs to be noted, although it was mentioned above, that once a sample is run though the MS and a RAW file is generated, a report can be generated showing the microbes in the sample. If new microbes are added to a data group, only the RAW file needs to be re-analyzed to determine if the new microbe or microbes are present in the sample. It is an easy step to examine some of the historical RAW files using the updated data groups to determine if a microbe was detected and identified. The coronaviruses sequences were downloaded from the NBCI. Some of these sequences were only weeks old when used to examine samples collected before and during the COVID-19 outbreak.

7.4 Detection and identification of viruses using MSP

A convergence of hardware using mass spectrometers and the software ABOid and faster computers and rapid additions to sequencing methods all contributed to the success of MSP/ABOid. The first success was for single microbes. Grown in a microbiology laboratory, bacteria were identified, collected, and prepared for the MS. Individual MS files were then analyzed using the ABOid software and the bacteria was confirmed. Additional efforts combined bacteria and repeated the procedure which then detected and identified multiple bacteria (Jabbour et al. 2010; Wick C. H., *Identifying Microbes by Mass Spectrometry Proteomics* 2014).

The detection and identification of viruses collected by honeybees is used as an example for detecting viruses using MSP methods (Wick 2021). Many virus species and strains have been collected by honeybees across the United States. They have been detected and identified using mass spectrometry proteomics and the ABOid software which utilize unique peptides and associated genetic sequences to determine identification. Nine viruses are paramount in all the averages ranging around 15 unique peptides in the national average. The highest averages include African swine fever virus, camelpox virus, cercopithecine, goatpox virus pellor, lumpy skin disease, monkeypox virus, sheeppox, swinepox virus, vaccinia virus, and variola virus.

Several hundred viruses collected by honeybees were identified. Of these viruses, 103 were detected frequently enough to make an average. Of these 103, 14 were identified as standing above all the rest as seen in Figure 7.2, the national average for viruses. Five viruses can be grouped as belonging to the Herpesviridae, eight belong to the Poxviridae and the African swine fever virus stands alone. For all the viruses detected at low levels we have three families of viruses that dominate, herpes, pox viruses, and African swine fever. It is beyond the scope of this book to analyze "why" these particular viruses occur, but the fact that they do occur is equally important.

Camelpox, monkeypox, and vaccinia viruses show greater numbers of peptides than *Cercopithecine herpesvirus 5*. As to the meaning of these findings, well it is clear that these viruses are being detected and identified. Finding more than 80 unique peptides for camelpox virus is interesting, as it is for monkeypox and vaccinia.

The three major groups and the high number of unique peptides found in these groups should not distract from the detection and identification of the other viruses. Any of these other viruses could become a detection of interest.

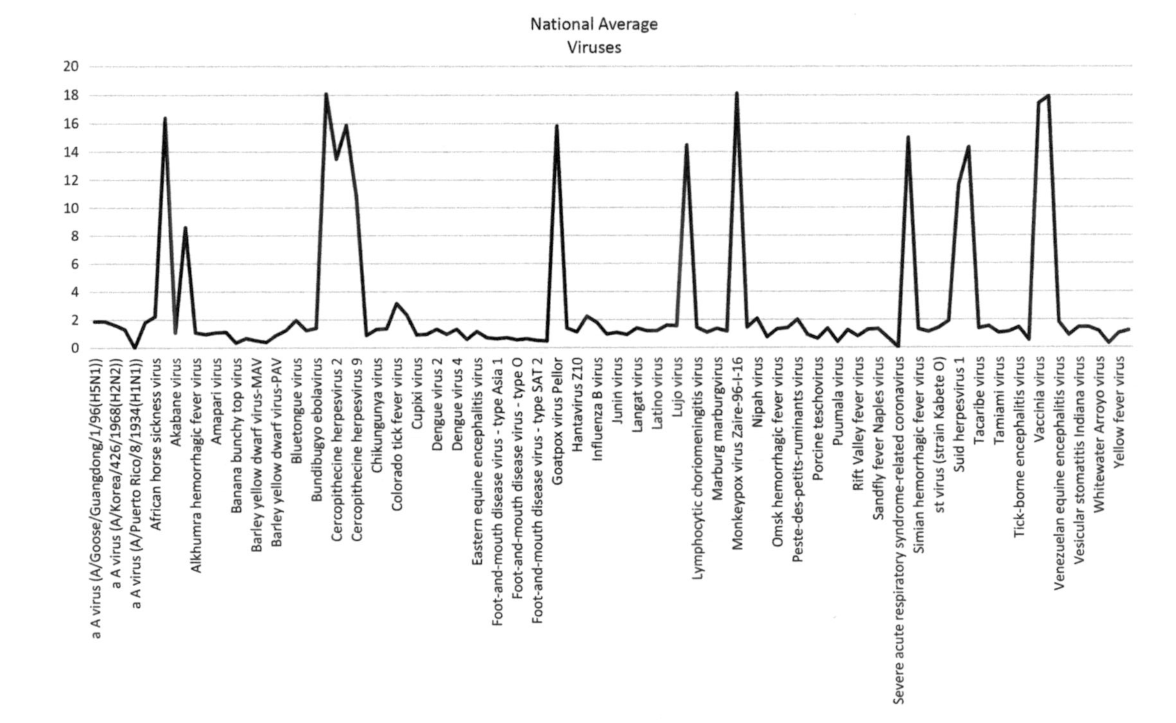
National Average
Viruses
20
18
16
14
12
10
8
6
4
2
0
a A virus (A/Goose/Guangdong/1/96(H5N1))
a A virus (A/Korea/426/1968(H2N2))
a A virus (A/Puerto Rico/8/1934(H1N1))
African horse sickness virus
Akabane virus
Alkhumra hemorrhagic fever virus
Amapari virus
Banana bunchy top virus
Barley yellow dwarf virus-MAV
Barley yellow dwarf virus-PAV
Bluetongue virus
Bundibugyo ebolavirus
Cercopithecine herpesvirus 2
Cercopithecine herpesvirus 9
Chikungunya virus
Colorado tick fever virus
Cupixi virus
Dengue virus 2
Dengue virus 4
Eastern equine encephalitis virus
Foot-and-mouth disease virus - type Asia 1
Foot-and-mouth disease virus - type O
Foot-and-mouth disease virus - type SAT 2
Goatpox virus Pellor
Hantavirus Z10
Influenza B virus
Junin virus
Langat virus
Latino virus
Lujo virus
Lymphocytic choriomeningitis virus
Marburg marburgvirus
Monkeypox virus Zaire-96-I-16
Nipah virus
Omsk hemorrhagic fever virus
Peste-des-petits-ruminants virus
Porcine teschovirus
Puumala virus
Rift Valley fever virus
Sandfly fever Naples virus
Severe acute respiratory syndrome-related coronavirus
Simian hemorrhagic fever virus
st virus (strain Kabete O)
Suid herpesvirus 1
Tacaribe virus
Tamiami virus
Tick-borne encephalitis virus
Vaccinia virus
Venezuelan equine encephalitis virus
Vesicular stomatitis Indiana virus
Whitewater Arroyo virus
Yellow fever virus

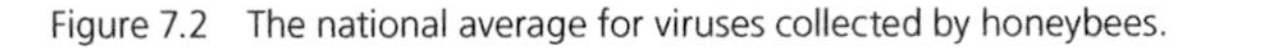
Figure 7.2 The national average for viruses collected by honeybees.

7.5 Examples of viruses detected by MSP methods using ABOid

The first thought when considering detection and identification is the wide variety of viruses detected. Figure 7.2 illustrates this diversity and gives a national average for the 103 viruses frequently detected.

Several other thoughts come to mind when considering these viruses. The main one is why? Why are we seeing these in our samples, and why are they so widely distributed? As you read below you will see that these groups have in common a wide distribution and a history of hardy survivability. This would indicate that the pox viruses, the herpesviruses, and evidently the African swine fever virus can survive in a wide distribution. All the regions show these viruses, there is no doubt they are here.

The final thoughts on this issue are two: "what do I do about them?" and "what does this mean to me?" First, these viruses have been around a long time. We have not sustained, as far as we know, any major negative side effects. We consume them, we breathe them, we live with them. The population has immunity from many microbes, and as a result they are not a present threat to the population. As further evidence, it appears that most microbes are not a direct threat to us because we have not sustained an outbreak of any of these viruses among the population. This means that although we have detected these viruses, not all of them are a threat to us and we should treat their detection not as a worry but as interesting information. The natural microbe population should be studied as there may be pathogens lurking. Tetanus, polio, common colds, and other microbes are in this microflora, and weakened or immunocompromised individuals may be at risk. While we could be seeing the natural occurring microflora that surrounds us all, it also is evident that we are detecting pathogenic microbes as well. Monitoring these pathogenic microbes that have the potential to become a threat is useful considering that some of these microbes may be just waiting to pounce if they have the chance. Outbreaks of infection that lead to a pandemic are a different matter. In these cases, such as influenza and COVID-19, it would be helpful to know where the infectious virus is located in the environment so it can be controlled and eliminated.

7.5.1 African swine fever virus (ASFV) – Variola porcina

African swine fever virus (ASFV) is the only species in the order Asfuvirales, family Asfarviridae, and genus *Asfiviru*. Variola porcina is a large DNA virus of the phylum Nucleocytoviricota. It is a large, icosahedral, double-stranded

DNA virus with a linear genome of 189 kilobases containing more than 180 genes (Linda K. Dixon 2008).

Although ASFV is not known to cause disease in humans it has been a problem in swine. It was thought to have evolved around 1700. It can be spread by ticks and pigs and also by food products that contain the virus. It causes hemorrhagic fever in pigs usually within a week of infection.

Given this information, a question then remains, why are we seeing it among the prominent viruses isolated from honeybees? This virus is present everywhere across the United States as evidenced by the analysis of honeybee collections. One thought on this is that following the widespread pig farming in the early 1900s and the first major outbreaks of ASFV at that time to sporadic outbreaks from time to time, the virus became endemic throughout much of Africa, Europe, and when it crossed the Atlantic in the Caribbean. In 2018 the virus spread to Asia. Since the virus can remain in an infectious state in the tick vector (*Ornithodoros*) for months or up to years it is likely that the virus has spread among the natural hosts, warthogs, bushpigs, and soft ticks. The continued spread of this virus since being first noticed to recent times would indicate a rationale for detecting ASFV. Much of this spread could have been through travel and the import and export of pigs and pig parts.

Also associated with pigs is Aujeszky's disease (also known as pseudorabies). This is a viral disease of pigs and endemic in most parts of the world. It is caused by *Suid Herpesvirus 1*, a member of the subfamily Alphaherpesvirinae and the family Herpesviridae. The virus infects a variety of mammals, but only pigs are able to survive a productive infection and are thereby considered the natural host (Kluge et al. 1999).

The national average for ASFV is 16.4 unique peptides (Figure 7.2). Twenty viruses stand out as having a greater average of unique peptides. These are listed below with a brief description as to their averages and descriptions.

7.5.2 *Alcelaphine herpesvirus 1* (AlHV-1)

Alcelaphine herpesvirus 1 (AlHV-1) is a large DNA virus of the phylum Peploviricota, also known as malignant catarrhal fever. *Alcelaphine herpesvirus 1* serves as the prototype virus of the *Macavirus* genus of the Gammaherpesvirinae family Herpesviridae. Bovine malignant catarrhal fever (BMCF) is a fatal lymphoproliferative disease (Toole & Li 2014). The national average for AlHV-1 is 8.6 unique peptides (Figure 7.2).

7.5.3 Camelpox virus (CMLV)

Camelpox virus is a large enveloped DNA virus that is taxonomically assigned to the family Poxviridae, subfamily Chordopoxvirinae, and genus *Orthopoxvirus*. Other members of the genus (i.e., the orthopoxviruses) include important human pathogens such as variola (smallpox), monkeypox, cowpox, and vaccinia viruses, in addition to those of lesser importance such as ectromelia, raccoonpox, skunkpox, taterapox, and volepox (Moss & Poxviridae 2013). Camelpox (CMLV) is a contagious viral disease of camels that occurs throughout the camel-breeding countries of northern Africa, the Middle East, and Asia (Balamurugan et al. 2013). Like other poxviruses, camelpox virions show a high degree of environmental stability and can remain infectious over several months (Rheinbaden et al. 2007). The national average for CMLV is 18.1 unique peptides (Figure 7.2).

7.5.4 *Cercopithecine herpesvirus 5* (CeHV-5)

African green monkey cytomegalovirus (CMV) and *Cercopithecine herpesvirus 5* (CeHV-5) are viruses in the genus *Cytomegalovirus,* subfamily Betaherpesvirinae, family Herpesviridae, and order Herpesvirales. African green monkeys (*Chlorocebus* spp.) serve as natural hosts. The national average for CMLV is 15.9 unique peptides (Figure 7.2).

7.5.5 Goatpox virus Pellor (GTPV)

Goatpox virus Pellor, also known as GTPV strain Pellor (PL), may affect goat and sheep populations. The virus is endemic in southwestern Asia, India, and northern and central Africa. GTPV is one of three recognized members of the genus *Capripoxvirus*. The national average for GTPV is 15.8 unique peptides (Figure 7.2).

7.5.6 Lumpy skin disease virus (LSDV)

Lumpy skin disease virus (LSDV) is a double-stranded DNA virus and a member of the *Capripoxvirus* genus of Poxviridae. Capripoxviruses (CaPVs) represent one of eight genera within the Chordopoxvirus (ChPV) subfamily including the sheeppox virus and goatpox virus. The capripoxviruses are brick-shaped and different than Orthopoxvirus virions in that they have a more oval profile and average in size between 260 nm and 320 nm (Tulman et al. 2001).

Lumpy skin disease (LSD) is an infectious disease in cattle. It has spread rapidly through the Middle East, southeast Europe, the Balkans, Caucasus,

Russia, and Kazakhstan (World Animal Health Information Database (WAHID) Interface 2020) and is considered to be one of the emerging threats to Europe and Asia (Machado et al. 2019; Allepuz et al. 2019). LSDV mainly affects cattle and is also seen in giraffes, water buffalo, and impalas and other animals. Fine-skinned *Bos taurus* cattle breeds such as Holstein-Friesian and Jersey are the most susceptible to the disease. The national average for LSD is 14.5 unique peptides (Figure 7.2).

7.5.7 Monkeypox virus Zaire-96-I-16 (MPV)

Monkeypox virus (MPV) is a double-stranded DNA, zoonotic virus and a species of the genus *Orthopoxvirus* in the family Poxviridae. It is one of the human orthopoxviruses that include variola (VARV), cowpox (CPX), and vaccinia (VACV) viruses. But MPV is not a direct ancestor to, nor a direct descendant of, the variola virus which causes smallpox (Breman 1979).

It is interesting that monkeypox is associated with monkeys but they are not the main reservoir of the virus. Antibodies have been found in a variety of animals (Khodakevich et al. 1986).

The MPV has been found in ground squirrels, and they may be a reservoir of the virus (Sergeev, et al., 2017). This would lead to the suggestion that MPV may be much more prevalent in the environment then commonly thought and that the honeybees are picking it up just as they would any other virus in their activity. The national average for MPV is 18.1 unique peptides (Figure 7.2).

7.5.8 Sheeppox virus

The national average for sheeppox virus (SPV) is 15 unique peptides see goatpox in Figure 7.2.

7.5.9 Vaccinia virus (VACV)

Vaccinia viruses are close relatives of the smallpox virus (VARV) and are also pathogenic to humans. These include the Old World orthopoxviruses, VACV, cowpox (CPXV), and monkeypox (MPXV). Rodents are the major natural reservoir of cowpox and monkeypox (Shchelkunov S.N., 2005) (Moss, Poxviridae: The viruses and their replication, 2007). The national average for Vaccinia is 14.5 unique peptides (Figure 7.2).

7.5.10 Variola virus (VARV)

Variola is a large brick-shaped virus measuring approximately 302–350 nanometers by 244–270 nm, with a single linear double stranded DNA genome 186 kilobase pairs (kbp) in size and containing a hairpin loop at each end (Roossinck 2016).

Smallpox is the disease caused by VARV, which belongs to the genus *Orthopoxvirus*. The declaration of its eradication from the human community by the World Health Organization (WHO) in 1980 did not, however, remove it from the list of potential and dangerous pathogens since the virus remains in laboratories for study and it has been known to reoccur throughout history. This possibility is furthered by the fact that people have not regularly been vaccinated against VAR for a long time and a susceptible and vulnerable population has grown, making a reemergence of the virus possible (Singh et al. 2012).

It has been detected at low levels in the environment, but still detected and at such a level to separate itself from the more than 500 other viruses found in the background. The national average for VARV is 14.5 unique peptides (Figure 7.2).

7.5.11 Discussion of viruses detected by MSP and ABOid

Hundreds of sequenced viruses have been identified. Many are common in the environment, and some have relationships with plants, and others are soil inhabitants; some are pathogenic. The national Average for viruses ranges from less than 2 to 18 unique peptides. Nine are discussed in detail because they stand out from the field by having 12–18 unique peptides.

The classification of viruses is ongoing, and some of them may be underspecified at the moment because many are not sequenced and remain un-named. Viruses have a high mutation rate, and the number of species or strains within a genus and species can rapidly change. The rapid increase in sequencing viruses will contribute to identification and cement phylogenic relationships.

Environmentalists are encouraged to study these viruses and compare them with the bacteria and fungi to determine mutual relationships for regions, associations with other data such as activities that affect a region, area, or a locality.

The number of virus species that can be identified is continuing to increase due to the increasing number of sequences added to the NCBI. More than 50,000 viruses have been sequenced; the number increases daily.

The ability to add new microbes has been discussed, but it remains important to remember that as new viruses are sequenced and added to the Virus

Data Group the samples that have been analyzed can be re-analyzed using the old computer files. In this manner there is no need to collect a new sample, and large archived sample sets can be re-examined and a search made for the new viruses. An example of this is the SARS and COVID-19 viruses. Recently over 450 new viruses have been added.

7.6 Adding new viruses using MSP

Figure 7.3 has five microbes. These microbes represent the microbes in an original data group. These microbes were identified and their sequences downloaded from the NCBI, and added to the original data group. A sample file (RAW) that has been processed through the MSP is then analyzed using the ABOid software. The five new microbes were identified as being of interest and their sequences downloaded from the NCBI and added to the original data group. The new data group now contains the original sequences and the new sequences for a total of ten microbes. Figure 7.4 has the ten new microbes in a re-analyzed result which contains the new microbes and the new results that demonstrate the process. In this manner numerous new microbes can be added as their sequences become available.

New microbe sequences are being added every day to the NCBI database. Considering that the number of sequenced microbes has increased from a few hundred to more tens of thousands, it can be seen that it is useful to add these new microbes to any detection and identification method.

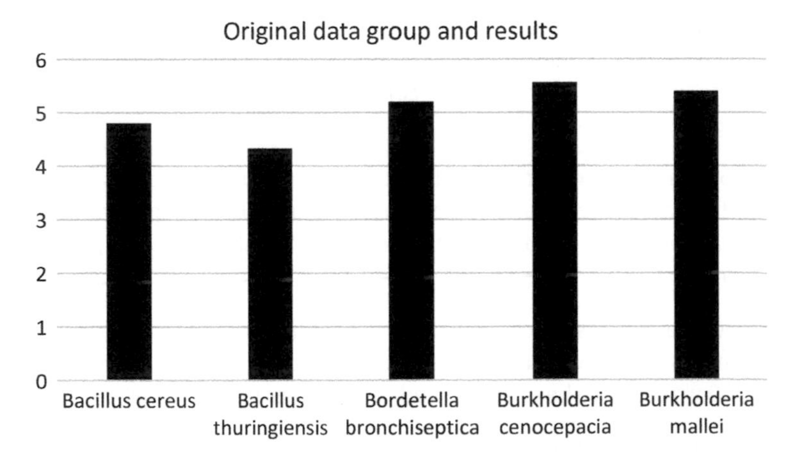

Figure 7.3 Example of how to add microbes to MSP methods.

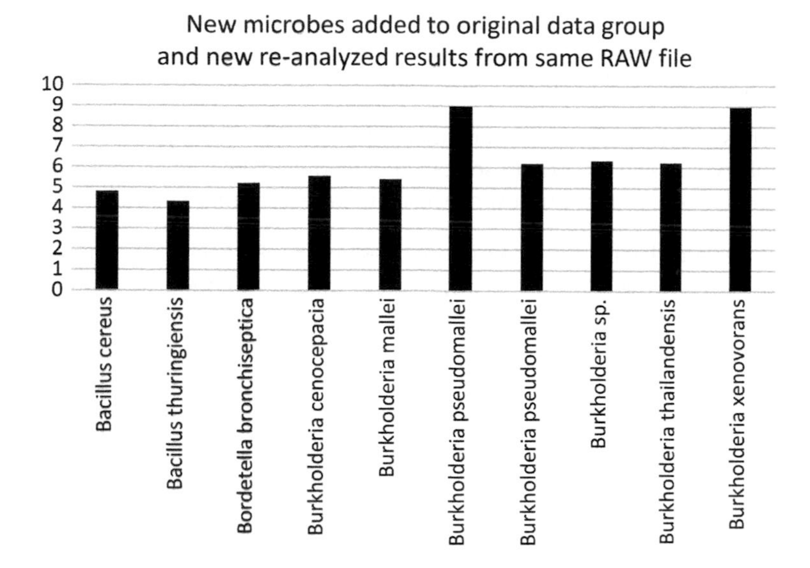

Figure 7.4 The new MSP has the new microbes in a re-analyzed result which contains the new microbes and the new results that demonstrate the process.

The original Virus Data Group used in the analysis of honeybee samples consisted of less than 100 viruses. This number was increased to 200, 300, and then 750 viruses by simply downloading new virus sequences and adding them to the Virus Data Group. This process can be accomplished in less than an hour moving at a deliberate and scientific speed, which makes the new data group more robust and useful. When the first 200 new virus sequences were added to the Virus Data Group it was interesting to see the new viruses show up in the MSP/ABOid analysis. The old charts could be compared with the new charts and it was straightforward to pick out the new detections and identifications. Sometimes there were several new viruses.

Bacteriophage sequences were added to the Virus Data Group, and the result was the ability to detect and identify many bacteriophages. This was important as a means to indirectly detect the associated bacteria from such plants as tomato, citrus, and other plants.

Likewise, microbes loaded into different data groups can be compared to look for those associations between fungi and bacteria, and viruses, etc.

This capability was exercised in looking at the new COVID-19 outbreak. When the sequence of the novel SARS virus isolated from Wuhan-Hu in China was available it was downloaded from the NCBI. It was not long afterwards that additional sequences were made and these were downloaded and added to the data group, and likewise until more than a hundred sequences of COVID-19 were added.

In this manner selections from more than 24,000 Eukaryotes, 437,000 Prokaryotes, and 50,000 viruses can be added until they are all added. This would make for a very robust capability to detect microorganisms.

Viruses are classified by their genetic sequences. Particle counting methods such as the IVDS use a physical ion-mobility method to count the individual viruses and separate them by the virus size (Wick C. H., 2015). By separating viruses by size, IVDS has the ability to give an approximation of the type of virus, usually classifying them into family groups. The major benefit of IVDS is that it is a rapid means to detect the presence of a virus in a sample, using size to classify it into a preliminary identification, for example, influenza is 92 nm or 102 nm in size depending on if it is type A or type B. Another benefit of IVDS is that since it is not restricted by chemical reactions, it can detect all the viruses in a sample at the same time. This feature makes IVDS a perfect method to screen samples and verify identification by another means, such as MSP/ABOid. Since most samples from the environment are typically negative (unless looking at a preselected population) for flu or COVID-19, IVDS is again indicated as a means to screen large numbers, only referring positive samples for verification.

All these earlier methods are useful, but it has become the standard to verify a detection based on a classification scheme based on DNA or RNA sequences. The resulting phylogenic relationships are useful in following genetic drift and mutations in a particular strain that result in a new strain for a particular microbe. This ability to mutate and create new strains of a microbe is responsible, in part, for many of the new sequences that we have seen over the last ten years.

Other reasons that contribute to the increase in the number of microbes sequenced are that there are probably trillions of microbes and fast sequencing instruments are being developed. These two reasons point to a method such as MSP/ABOID for detection and identification simply because new sequences can be added quickly and just as quickly be placed into operation and the new microbes detected in samples. This capability of MSP/ABOID is also useful because since it is software it can re-analyze old files and determine if any of the newly added microbe sequences lead to a new detection and identification.

7.7 Coronavirus detection including SARS

In December 2019 a new severe acute respiratory syndrome (SARS) coronavirus was identified as infecting humans. Specifically, this particular severe acute respiratory syndrome coronavirus 2 (SARS-CoV-2) was found to be a novel coronavirus called "SARS-CoV-2" (previously referred to as 2019-nCoV), and it was determined to be a new strain that had not been previously identified in humans. The disease that is caused by SARS-CoV-2 is called "COVID-19". COVID-19 stands for corona (CO) virus (VI) disease (D) and 19 (2019), the year that the virus was detected.

COVID-19 like other viruses has rapidly mutated to form different strains of the virus. For example: the strain in western Canada originated in Iran, as did the strain in New Zealand and Australia. The Iranian line originally came from China as did some infections in Australia. There are European pockets of the virus from China. COVID-19 arrived from Italy into South America and Mexico as did many of the UK infections. Some strains may have passed through the Netherlands and Belgium before arriving in the UK. Needless to say, COVID-19 like other viruses tends to move around as people move around, and mutates and produces new strains. During the beginning of the COVID-19 outbreak there were only a few (15) sequences for the strains of COVID-19, and this quickly grew to nearly 100 sequences in 2 months and continues to expand, not unlike what was seen with the H1N1 influenza outbreak in 2008 (Wick 2009).

In this application the sequence for severe acute respiratory syndrome coronavirus 2 isolate Wuhan-Hu-1, complete genome – GenBank: MN908947.3, was downloaded from the NCBI and put into the SARS Data Group and used for the analysis of honeybee samples. It was quickly discovered as many more strains were being sequenced that one download was not going to be sufficient, and over 100 sequences were downloaded and integrated into the SARS Data Group. The SARS Data Group was then further augmented to include all the complete sequences for other coronaviruses, and renamed the Coronavirus Data Group with over 250 coronaviruses.

7.7.1 Coronaviruses

Aside from the 93 SARS or COVID-19 coronaviruses, there are many others of interest. Several bat coronaviruses exist in nature, as well as bovine, goose, duck, feline, rat, sparrow, swine, and turkey. The sequences for these coronaviruses have all been downloaded, added to the Coronavirus Data Group, and used to analyze the honeybee samples. A national average was determined (Figure 7.5).

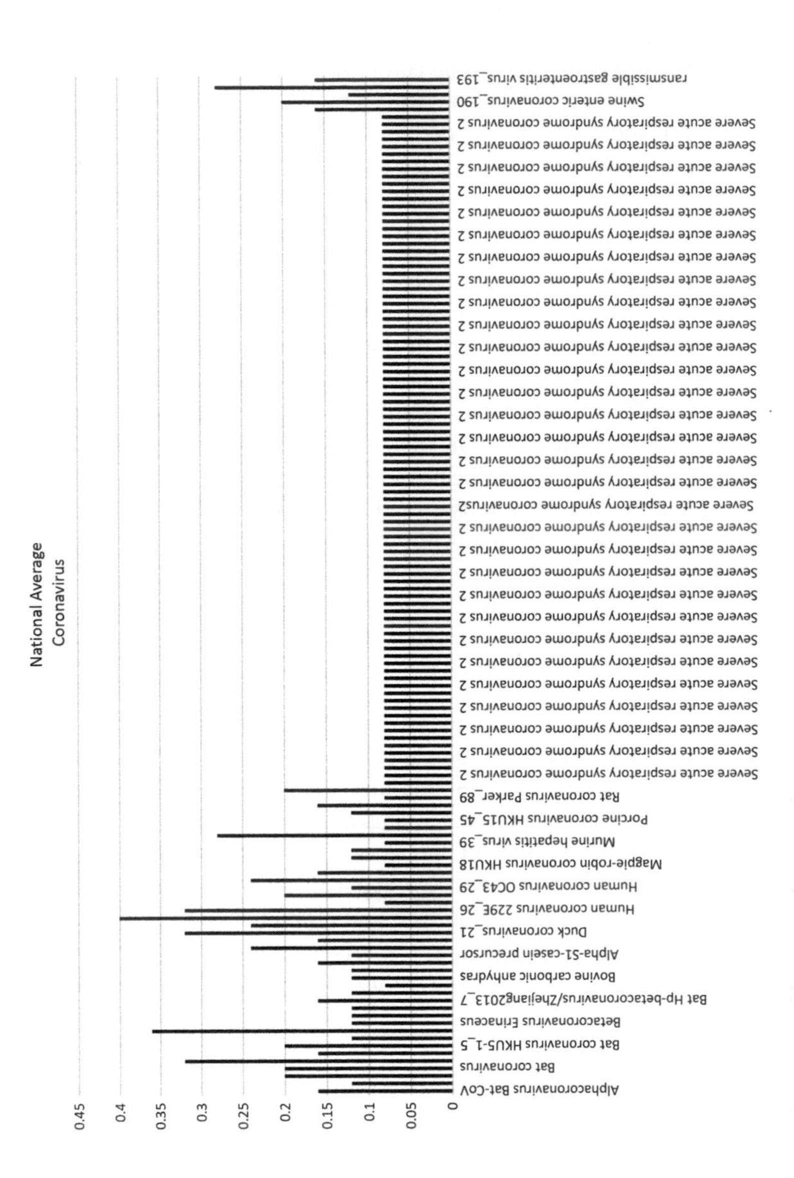

Figure 7.5 National unique peptide averages for coronavirus.

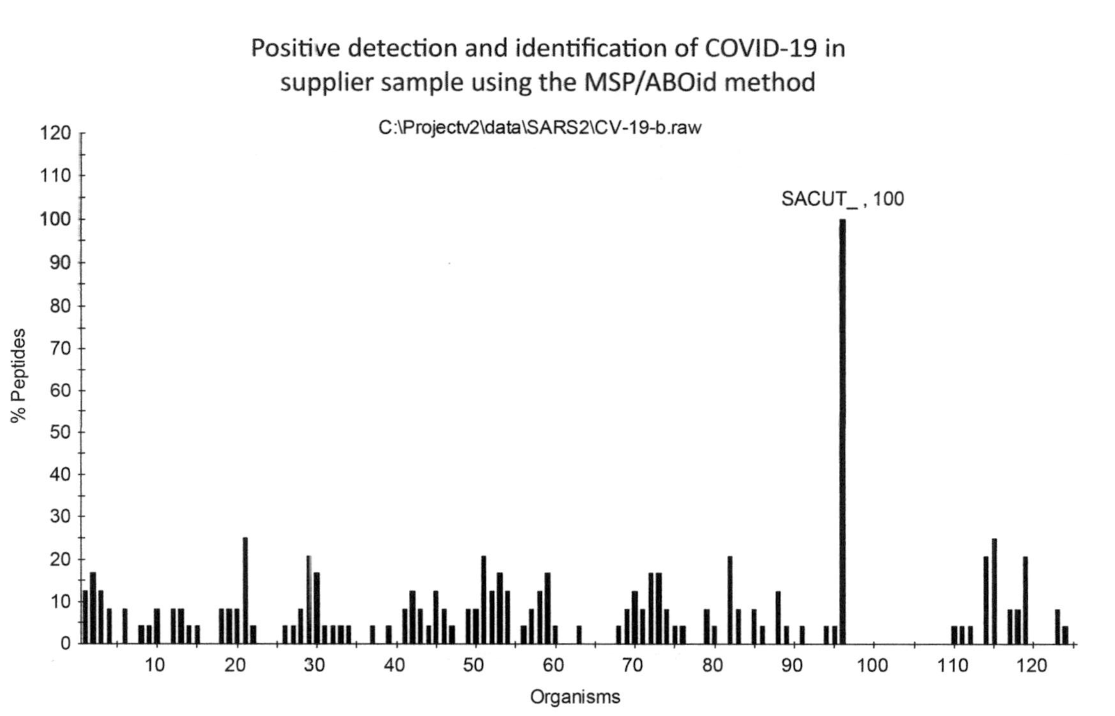

Figure 7.6 Positive detection and identification of a sample of COVID-19 from a supplier.

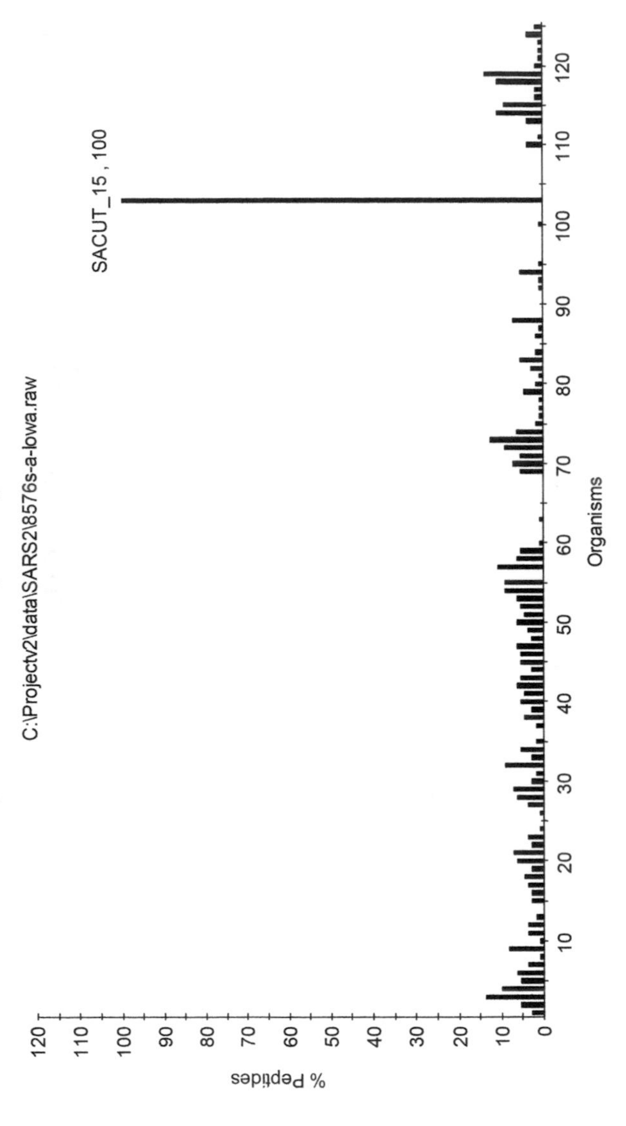

Figure 7.7 Honeybee smoothie sample with a detection and identification of COVID-19.

7.7.2 National average for coronavirus

7.7.2.1 Verifying COVID-19 detection

Figure 7.6 illustrates the detection and identification of a lab-supplied COVID-19 sample. The main peak is the virus, the smaller peaks are other items not of interest. This is a clear result of the MSP/ABOid analysis using the standard protocols.

Figure 7.7 is a honeybee sample showing the positive detection and identification of COVID-19. The main peak is the virus, the smaller peaks are other items not of interest. This is a clear result of the MSP/ABOid analysis using the standard protocols for all the honeybee samples.

7.7.2.2 COVID-19 detection discussion

This is of interest for many reasons other than demonstrating that coronaviruses can be detected from honeybees or rather that the honeybees collect many microbes from the environment and among those microbes are the coronaviruses, including COVID-19. The MSP/ABOid analysis using the standard protocols has shown to be successful in detecting a wide range of microbes, this just being one group. The distribution of the coronaviruses among different regions illustrated that not all the regions are the same and that they are different in type of virus and their frequency.

It was considered important to demonstrate that the MSP/ABOid analysis using the standard protocols could detect and identify COVID-19. By taking a COVID-19 sample supplied by a laboratory and comparing it directly with a sample prepared from a honeybee sample, it is demonstrated that the MSP/ABOid analysis is accurate.

The detection of a variety of other coronavirus strains is interesting. It appears that strains of human coronavirus and infectious bronchitis virus in several of the regions might be useful to monitor as well as COVID-19. The spread of a virus such as COVID-19 could prove valuable in determining the spread and the relative averages for each area of interest, such as a town or state.

7.8 Summary of MSP-ABOid detection of viruses

- Detects multiple viruses in single sample
- Identifies viruses to strain level
- Uses the genetic sequence as downloaded from the NCBI
- Can detect thousands of viruses in a single sample

8
Discussion

Four major methods are used in virus detection in historical order, electron microscopy (EM), molecular (PCR and antibody), physical counting (IVDS), and mass spectrometry proteomics (MSP). Each method has strengths and also weaknesses. Often the best solution is to use a combination of methods. Suggestions are made on the use of these methods to best enable the desired conditions for the detection of a virus under different scenarios. Ideas are presented for the protection of important people, small groups of people, and larger groups. The need for accuracy in the detection as well as speed to be able to detect is discussed. Methods are discussed with regard to their affordability. The chapter discusses which methods are best indicated for detecting unknown or un-sequenced viruses as well as multiple viruses or a combination of unknowns and multiple viruses in a single sample. Early warning of viral outbreaks is important in some situations; the methods used in clinics and hospitals might be different than methods used for environmental samples, agriculture, animals (mad cow disease and foot-and-mouth disease come to mind), and insects (the detection of viruses in honeybees is important). The optimized solution for each use is discussed. The take away from this book is that we can detect viruses.

8.1 Introduction

Detecting viruses from the biological soup that we live in which contains trillions and trillions of microbes (the mixture of fungi, bacteria, and viruses) is an immense task. Science has made incredible advances in the last hundred years when we first suspected a particle smaller than bacteria (a virus-like particle) existed beyond our ability to visualize it with photonic (light-based) microscopes or otherwise sort it out. The advancements in the 1930s, 1990s, and early 2000s allowed the visualization, phylogenic mapping, and the current sorting out of virus by family, species, and strain. New methods were invented as technology improved in computing, biomolecular methods, particle counting, and mass spectroscopy. The strengths and weaknesses of each of these methods need to be considered when detecting viruses under different conditions. In Figure 8.1 some of

DOI: 10.1201/9781003106623-8

Virus Detection Methods

General Requirements	Electron Microscopy	PCR	Antibody	IVDS	MSP
Proven Science	✓	✓	✓	✓	✓
Accurate Detection	✓	✓	✓	✓	✓
Affordable		✓	✓	✓	
Reagents	✓	✓	✓		✓
Detection for Unknown Viruses	✓			✓	
Detection for Multiple Viruses	✓			✓	✓
Fast Results		✓	✓	✓	

Figure 8.1 Differences in types of virus detection methods.

the general requirements for virus detection are considered. These are basic requirements and do not include all the capabilities for each technology as many have multiple applications and uses. When considering the use as a virus detector certain capabilities are more important depending on the application. In the following charts these requirements are customized; for example, if the application for detecting viruses needs to be rapid, that feature becomes important; if you have the time, it is less important. The check marks are used to indicate a "yes" answer; however, there are situations where the consideration of other features becomes paramount, such as the need for special skill to operate or the need to create a new primer or the cost. The check marks are an indication and assume that the application is ready to run and many of the set-up requirements are already complete. Another example is that, although PCR is listed as affordable, this may change when the need to add in new primers and all the work that is required to get ready are considered. Likewise, electron microscopy, IVDS, and MSP methods require costly equipment and skilled operators. These methods need to be considered along with their benefits, such the ability to detect unknown viruses using EM and IVDS but not molecular or MSP methods. If you have the equipment then consider the operational aspects. It should be noted that most users of a particular technology are fervent supporters of their particular method and often claim that they have the solution for every situation. Clearly a combination of technologies is indicated due to wide and various requirements.

8.1.1 Proven science

All the main technologies have a long track record of proven science. Electron microscopy, both transmission (TEM) and scanning (SEM), has given wonderful pictures of viruses. These pictures have been useful in making the artist concepts (graphic illustrations) of the various shapes, sizes, and their attached proteins on their surfaces. The "spike" protein on the coronaviruses is interesting. Biochemical approaches likewise have given us detection methods based on genetics, and the resulting phylogenomic associations have helped to classify the viruses according to their genetic relationships. This has the advantage of sorting the 50,000 sequenced viruses and gives a path forward for all of the millions if not trillions of viruses yet to be discovered and sequenced. Discussion sometimes occurs around the identification of new strains of viruses as they are often closely related to other strains. It does, however, put the new virus into a proper relationship with the other viruses. The discoveries have continued making the technologies and their methods faster and easier to use, more reliable, and robust. The immunological approaches are similar. They are well-established approaches and are good for detecting the antibodies to a particular virus. They can also detect an unknown virus by pairing the antibodies discovered with the virus. Forty plus years of extensive experience has resulted in a wide variety of test kits and methods for detecting common viruses. They have limitations such as being able to detect only one virus at a time, although techniques such as multiplex arrangements add several single detection processes together in order to detect several viruses. Direct virus counting was invented next in the late 1990s, and the IVDS methodology is the common instrument. This process allows the direct counting of whole or intact virus particles and separates them by their size, small viruses (polio) to large viruses (smallpox), as confirmed by electron microscopy. IVDS can also determine the concentration of viruses. With this method and with a 4 nm separation, all viruses can be sorted by size and detected in one sample. It has to ability to detect any virus, sequenced or not. It does not identify them genetically or put them into a phylogenetic relationship. Finally, in the early 2000s, mass spectrometer proteomics (MSP) with the ABOid software was developed to a point where peptides could be identified and computers improved to the point where large computer codes such as ABOid could be processed. Combined with increases in the number of viruses sequenced all the known 50,000 viruses sequenced could be detected and identified in samples without the reagents needed for biochemical means. All the methods can detect viruses, the question is which method works best for a particular situation given their strengths and weaknesses.

8.1.2 Accurate detection

When considering the accuracy of detection all the methods have different degrees of accuracy and they need to be considered based on what kind of sample is being analyzed. When electron microscopy sees a virus, particularly one with distinguishing features, it has high certainty that the virus is present and some estimate as to the identity. Sample preparation is important, and some preparations may limit the accuracy. Likewise, when PCR and immunological methods are positive for a certain gene fragment or antibody to a virus, the method has high certainty that a virus is or was present. Contaminations within these samples can affect results leading to a lower accuracy. Some test kits are only 60–70% accurate for some samples. Direct counting using IVDS acts as a particle counter; it simply counts the virus particle with high accuracy. It also counts any virus or virus fragment. Large viruses (larger than 200 nm) can be disassociated by rough handling. The latest method, MSP, can detect and identify a wide range of viruses with high accuracy. Careful preparation of the sample and a certain amount of cleanliness is indicated to avoid background noise.

8.1.3 Affordable

The methods can be ranked according to cost per test: IVDS, immunology, PCR, MSP, electron microscopy. If they are ranked according life cycle costs (equipment is already acquired). Estimating the cost of the support equipment and supplies such as primers, reagents, etc., the ranking would change the list to: IVDS, PCR/immunological, MSP, and electron microscopy. Instruments and maintenance figure as large one-time purchases, such as electron microscopes, IVDS, and PCR machines. Likewise, reagents and manpower are expensive for electron microscopy, PCR, immunological methods, and MSP.

Another way to look at this is that once the cost of an instrument is made and the primers and reagents are made/purchased then the way to consider detecting a virus is by costs per sample over the long term. Even this consideration is tempered by the number of instruments available or as is the case for PCR/immunological the cost to produce a kit for testing. In cases where an expensive instrument is needed, once the detection instrument is available for use the detection of a virus or viruses can continue uninterrupted for an indefinite time, and the cost per sample decreases. Also, consider the detection of an unknown (not sequenced) virus or a man-made virus. Using PCR and immunological methods, it would soon be impossible to make test kits for all the 50,000 sequenced viruses or the trillions of unknown let alone man-made viruses. Immediate detection of an unknown or man-made virus presents another problem. Remember the n+1 rule.

One more thought: instruments can be used at any time and shelf-life is usually not an issue. Electron microscopes, IVDS, and mass spectrometer proteomics (MSP) can sit unused until needed, with no expiration date or shelf-life. There are some considerations for supplies for preparing samples for the EM and the MS. There are no limitations for IVDS; it can sit for a long time, maybe even years in a storage configuration and be brought to use quickly when needed. PCR and immunological methods have concerns with shelf-life (usually 18 months) and special reagents that are needed to fully configure this method of testing. One example of this is the lag time or delay in testing for a new virus and then the shelf-life of the testing kits. The kits are usually specific for a virus and need to be replenished or a new kit developed for a different virus. Reagents are usually best used "fresh". It is not unusual for a month to pass while the PCR and immunological supplies are readied to verify a detection made by IVDS or the mass spectrometer or even the electron microscopes. In this manner PCR and immunological methods are not recommended for screening samples; use the IVDS to detect and the mass spectrometer to confirm, and use PCR/immunological means later. One such use is in testing large populations for a particular virus or viruses when the virus is well characterized. Large numbers of kits can be prepared for this use. One main issue with kits is that they are a "one-time-use" and you can run out of kits. Instruments such as the IVDS can screen samples for years without a worry about running out of kits. The issue is that instruments such as IVDS are not numerous and cannot, at this time, reach a lot of people. EM, IVDS, and MSP are also limited by the number of samples they can test each day. One way around this limitation is to manufacture a lot of instruments and put them in strategic sites to be used when needed, and in the case of IVDS test a large number of people, and test again, and again as it is a multiple use technology.

Considering that all these technologies (methods) can detect viruses, the actual affordability is governed by how urgent the need is, how fast a detection is required, how many people are to be tested (maybe surfaces and things as well), how accurate, and how reliable. We need to consider: what are we trying to do, and how much can we budget? Sometimes a phased solution is indicated: fast partially accurate methods followed by highly accurate methods.

A long-term and affordable solution may be a network of rapid detection methods, backed up by highly accurate methods. A network that can stand by and be brought into use when needed, one that can detect and identify future outbreaks. This may be the most affordable solution.

8.1.4 Screens for unknown viruses

Only the IVDS and electron microscopy can detect unknown viruses at the moment. IVDS is a virus counter and it does not depend on what kind of virus it is counting. Electron microscopy is looking for the virus and either it sees or it does not. Other methods such as PCR, immunological, and mass spectrometer methods can mimic this by the detection to a higher classification such as at the family or genus level, but these methods are still limited to the list of sequenced organisms. If the new or unknown virus falls outside of this group it will not be detected and is invisible to genetic classification.

8.1.5 Ability to detect multiple viruses

IVDS, electron microscopy, and MSP all can detect multiple viruses in a single sample. The MSP method is limited to those viruses that are sequenced and ready for detection; likewise, PCR and immunology methods can multiplex single detection methods into a larger array to detect multiple viruses. The multiplex is always limited to the virus plus one rule. There is always one more virus to detect, and in some cases that is the one that is important. It is important to consider the differences between these methods. The common features are that they both use sequenced information and both are as accurate as the genetics allow. There are also differences in how they function; MSP is an instrument that detects peptides and uses software to identify them and then sorts unique peptides to identify the virus. PCR and immunological methods use reagents, wet chemistry, and biomolecular methods to provide a means of detection. MSP can run continually, given adequate supplies, at a constant rate; PCR can test until it runs out of kits or a new virus emerges that is not detected by the kits.

8.1.6 Quick results (5–10 min)

Ranking the four methods as to how fast to run a sample and obtain results yields: IVDS (less than 5 minutes), PCR (minutes to days)/immunological (minutes), MSP (hours to days), and electron microscopy (hours to days). That is assuming the methods are ready to test, are in-place, and that the sample is just being analyzed. Again, each technology brings a slightly different approach to the problem.

Another consideration is the ability to share the information electronically. This implies a couple of methods. All the instruments produce a file with either an image (electron microscopy), a file of the analysis (PCR and immunology, IVDS), or a file that can be reanalyzed (MSP). The first three methods can share results, but the analysis needs to be re-run to provide updated information. The file of the MSP method contains all the biological information, even the information for un-sequenced viruses, and can be re-run if, for example, new sequences are available for a new virus. This is done with software, and the MSP instrument is not needed. This is very useful if samples were analyzed around the world and sent electronically to a central site for various purposes. In this manner archived samples could be re-examined as the virus sequences are updated. It should be noted that the number of viruses sequenced has increased by 10,000 in the last few years. The number of viruses may number in the trillions, and new sequences are added on a continual basis (Figure 8.2).

Virus Detection Methods

Requirements for Clinical Virus Detection	Electron Microscopy	PCR	Antibody	IVDS	MSP
Rapid Detection		✓	✓	✓	
Multiple Testing Centers				✓	
Reusable	✓			✓	✓
Affordable		✓	✓	✓	
Minimum Logistics		✓	✓	✓	
Detection for Unknown Viruses	✓			✓	
Detection for Multiple Viruses	✓			✓	✓

Figure 8.2 Challenges of virus detection methods.

8.2 What are the challenges when detecting viruses?

8.2.1 Interference

Each of the four methods has its own unique issues that cause interference. Again, in historical order, electron microscopes, both transmission and scanning, have concerns with the preparations of the sample. In some cases, more than 300 different methods have been used to prepare a sample before artifacts were controlled. Cutting thin slices for transmission microscopes can frequently be a factor, and there are many other issues; see Chapter 4 for more details. PCR and immunological methods can have interference from heavy metals and complicated sample matrices; see Chapter 5 for more details. IVDS methods have the least interference. The "Virus Window" limits interference from other particles, and the sample matrix does not appear to be an issue; see Chapter 6 for more details. Mass spectrometry has issues with being very sensitive and picking up ambient contamination; see Chapter 7 for more details. A solution to this issue is to use multiple means to detect a virus, thus maximizing the chances of detecting a virus.

8.2.2 Sensitivity or trust in a particular technology

This is the question of a false negative. If the detection method cannot detect a virus because there are too few of them, or they are not equally dispersed in a sample, or simply they are missed in collection, then the virus is not detected and the test shows a negative when in fact it is a positive. Adherents of each technology will frequently loudly proclaim different levels of detection, e.g., the theoretical limit, and not the actual level of detection, with lots of charts and arrows to prove their point. There is a functional level of detection that gives a reasonable confidence in detection. This should be considered when the number of viruses needed for an infection is determined. It is different for each of the viruses. People need to be cautious when selecting a virus detector, and due to the need to safely test high value groups using more than one method should be considered. Probably the most reassuring is the electron microscope because it will actually see the virus. Still, no image does not always mean no virus.

High value groups should consider using the IVDS for early detection, followed by a multiplexed PCR method, followed by MSP and then electron microscopy. The reason for this grouping is the speed of detection, fast to slow. To conserve resources, test for a virus using a reliable re-usable method. Since it can be expected that most samples will be negative it makes sense to test only the positive detections to confirm. There is no point in using a kit for every test when most are negative, you would run out of kits. This is followed by the MSP and a final test by electron microscopy. As the population to be tested increases this procedure can be scaled. Mass populations remain a challenge. Clearly, to test a mass population with a kit that has at the best 70% accuracy leaves a large number to continue to infect, thus complicating control.

8.2.3 Other things

One concern is the reliance on a single technology for detecting viruses and protecting high value groups. Likewise, users need to be aware of the limitations of each technology and use them accordingly (Figure 8.3).

8.3 Clinical

8.3.1 Clinic

Usually in a clinic there is 5–10 minutes to test a sample for various requirements. A1C tests for glucose usually takes a few minutes, whereas blood panels take a little longer (days). Testing for a virus or viruses could be made upon entering a clinic or while waiting by IVDS. If positive, then test with an antibody kit or PCR for a specific virus and if negative by MSP.

Virus Detection Methods

Requirements for Environmental Virus Detection	Electron Microscopy	PCR	Antibody	IVDS	MSP
Portable Technology		✓	✓	✓	
Detection for Multiple Microbes	✓				✓
Reusable	✓			✓	✓
Affordable		✓	✓	✓	
Minimum Logistics		✓	✓	✓	
Detection for Unknown Viruses	✓		✓	✓	
Detection for Multiple Viruses	✓			✓	✓

Figure 8.3 Considerations of using virus detection methods for clinic use.

8.3.2 Centralized testing center – separate from a hospital

Such an operation that routinely processes samples for diagnostics could easily test for viruses. Again, what is indicated is a method for testing for the presence of a virus (IVDS) followed by immunological tests for a specific virus followed by PCR and MSP if the IVDS test is positive.. Initial screening might take a few minutes while verification and identification may take longer depending on the virus. Common viruses such as influenza, hepatitis, and others have well-developed PCR and immunological tests for verification. Unknown viruses or less well-known virus may take longer to process.

8.3.3 Hospital

When you arrive at a hospital, the technique of asking people questions and having their temperature taken in an attempt to screen out COVID-19 is marginal. Accurate, rapid detection is indicated. A reusable technology, an inexpensive technology, and one with minimum logistics is needed. It is important to screen for multiple viruses and unknowns as the target virus may change. Screen for a virus and then confirm identity with a second technology if needed.

Other requirements could utilize centralized testing. This practice would be for the screening of staff and others as indicated. It is important to consider the false positive and false negative results; it is desirable to have these occurrences as low as possible. It is important to have an accurate indication of a negative and an accurate detection. The indicated order of use is: IVDS followed if a virus or viruses are detected by PCR or immunological test for a particular virus or by MSP for multiple viruses to gain an identity (Figure 8.4).

8.4 Environmental

Viruses are found in nearly every ecosystem. They are associated with all animals, all plants, and probably with every other living organism. As a result, virus detectors are indicated for all people working in these areas. From agriculture to zoology there is a need for a method to identify and classify viruses. Let us consider a few examples.

8.4.1 Agriculture – insects (bees), plants, and animals

Domestic and wild animals have viruses. Some among domestic animals are economically important, such as foot-and-mouth disease. This is a hardy environmental virus important to cattle. Roughly 20 nm in size, it is a simple virus to detect. It is frequently in an environmentally challenging matrix – the barnyard. If you are trying to detect this virus in a natural setting use

Virus Detection Methods

Requirements for Pandemic Virus Detection	Electron Microscopy	PCR	Antibody	IVDS	MSP
Portable Technology		✓	✓	✓	
Detection for Multiple Microbes	✓				✓
Reusable	✓			✓	✓
Affordable		✓	✓	✓	
Minimum Logistics		✓	✓	✓	
Detection for Unknown Viruses	✓			✓	
Detection for Multiple Viruses	✓			✓	✓

Figure 8.4 Use of virus detection methods for environmental samples.

IVDS to detect followed by other methods to confirm or identify. Most PCR and immunological methods find backgrounds such as the barnyard clutter challenging, often interfering with an accurate detection (Chapter 5). Instruments such as the IVDS, MSP, and electron microscopes do not have this limitation. IVDS can detect all the viruses in the matrix (Chapter 6). MSP is indicated for the identification but takes more time for an analysis (Chapter 7). Electron microscopy is useful for detailed study, but takes more time (Chapter 4). Depending upon the time constraints of the virus detection and mobility of the testing methods, different means can be used to detect viruses. If the number (concentration) of viruses is to be determined, or the identification is needed or the relationships of the virus to the background matrix or a combination of these requirements are needed a technology or combination of technologies can be used.

Likewise, insects frequently carry or harbor viruses. This is particularly the case with the honeybee which is of particular interest because of its importance as a pollinator and in producing honey. Several, sometimes closely related, viruses cause disease in the honeybee (see Chapter 7 for examples). Variola mites have their own viruses, and these mites are in turn related back to problems with the honeybee. Countless other insects have their own viruses which frequently cause disease. The bark beetle is an example. The study of all these viruses and their insect hosts are important for several reasons including their economic impact.

8.4.2 Water

Water is essential and needs careful monitoring. Water quality is frequently tested, and testing for microbes is often routine. Testing for viruses is less routine, because of rough water treatments which are supposed to remove most microorganisms. Many viruses can enter the water system through a wide range of conditions, and methods for testing for common gut viruses are indicated. Routine testing of water usually has the luxury of having time to test and any technology can be used. Again, a combination is recommended simply because of the limitations. It is suggested to test for multiple viruses.

Water is among the easiest of the sample matrices encountered. For this reason, instruments such as IVDS have a fast sample throughput. PCR and immunological methods have minimal interferences, and MSP and electron microscopy have simple preparations. It should be mentioned that water can dilute the number of viruses per sample, and metabolize molecular components.

Naturally occurring water may have lots of viruses. IVDS has detected numerous viruses in various sea water samples, for example. Streams and well water most likely have their share of viruses.

8.4.3 Research

Since viruses are nearly omnipresent it makes sense to study them. There are more than 50,000 viruses that have been sequenced, and the number is growing. As we understand the relationships between fungi, bacteria, viruses, plants, and animals we may gain insight into the part they play in the world. Viruses cause a host of diseases. That alone is a vast body of research to be done. How do we control viruses? Should we control them? One area that is important – that we should seek to understand them as viruses are a large part of the world in which we live.

Another area of research is how long do viruses survive? This has been a question asked since they were first recognized. A sample was collected and characterized and quite by accident left on the work bench. Out of curiosity the sample was retested and the virus was found intact and very happy. Bench top, hot areas, cold areas, dry areas, etc. are different, conditions were viruses are found and known survived. Follow-up investigations

Virus Detection Methods

Requirements for Other Virus Detection	Electron Microscopy	PCR	Antibody	IVDS	MSP
Technology for Public Use				✓	
Detection for Multiple Microbes	✓				✓
Reusable	✓			✓	✓
Affordable		✓	✓	✓	
Minimum Logistics				✓	
Detection for Unknown Viruses	✓			✓	
Detection for Multiple Viruses	✓			✓	✓

Figure 8.5 Using virus detection methods during a pandemic.

showed that this particular virus, a small RNA virus, survived for weeks, and weeks on the bench top, it is still there and being tested weekly. It was counted by IVDS and is being confirmed. This is contrary to what PCR teaches. This story continues, but continuing research is just beginning (Figure 8.5).

8.5 Use during a pandemic

It appears that society is slow to set up defenses against something they cannot see, or perceive as a threat, or simply do not want to think about. This problem may be generational; the current people appear not to be trained, they are not veterans and appear to be inexperienced. Also, the population is not trained. We have had experience with outbreaks, the influenza outbreak in 1918, and again in 2003, and it just floats around out there to repeat an infection.. The COVID-19 outbreak in 2019 and the variants are an example and so on with other viruses. We appear not to get ready for an outbreak until it is upon us and then we react. When we react to an outbreak, we frequently try the same old methods used in the past and appear not to be ready. This problem has been gamed (modeled by computer) for years, movies made, and little or nothing done right until after the immediate crisis is over. We need to get ready as we may not have the time to react next time.

Several viral diseases have been successfully combated and appear to be under control, for example Mumps, rubella, and the great success with smallpox. Others are maintained but still a problem; yellow fever and dengue fever are still among us to name just a couple. Some groups think that a vaccine for everything is the answer. This takes time to manufacture. Some people think it will not affect them and ignore safety protocols. We have seen this first hand. We need a network of virus detectors around the country and maybe even our foreign outposts, to give us early warning. Consider this:

A man-made virus is created and is a new and present threat and danger with a high mortality rate. It may appear as an unknown virus and certainly will not be publicly sequenced and posted. Do we accept another million casualties while we react to it? A better system needs to be prepared to give early warning and give time to react.

Develop an early warning network. This network needs the following capabilities – first, it has to have unlimited capability and not focus on just one or two viruses. Remember the virus plus one rule. Second, it has to be dependable and fast. It should be a system that is re-usable and have a low logistical footprint. It should be able to be stored, or in standby mode until needed. It should be easy to use. After a first detection, follow-up devices need to be available to confirm. Depending on the new virus which has already been detected and confirmed, systems can be brought to bear to classify it. Put it into a phylogenic tree and identify it, if possible. Then look at what can be done, vaccines, treatments, isolation, etc.

What technologies are then indicated? Looking at the list, it must be accurate and re-usable, and have unilateral detection, low logistics, and no shelf-life. Use the IVDS instrument to look for viruses, it meets all the requirements. It does not give the phylogenic relationship, but the MSP can be used and then an electron microscope can be used to show the virus for confirmation. PCR and immunological methods can then be developed to test for the virus.

This network can then give an early warning of an outbreak and provide a quick path to follow the course of the new virus and allow for full mobilization to contain and control it.

8.6 Other uses

8.6.1 Public use

The public use of a virus detector is one of the most challenging scenarios. Developing a test for general use has to take into consideration the wide variation in people and what they may test. Depending on the purpose and how it will be used and to minimize the down time we must consider the available technologies. A vast number of the public would misuse a testing procedure and minimize the effectiveness of a public testing system; for example a public kiosk in a retail store would need to be robust and durable and would likely be subject to damage. One thought is using an instrument like the IVDS to test for the presence or absence of a virus and then follow up with a more comprehensive test if positive. Such an instrument, although it can be operated by a high school graduate, could be subject to clogging and down time. People would not follow directions and test all sorts of sample matrices. A curious person may try to evaluate motor oil or other such compounds just to see what is detected. Such use by one or two people out of thousands would deny the virus detection to the rest. A way around this is to have someone at the point of use actually process that sample. For routine saliva samples a test could be performed in a few minutes. Likewise, the other technologies. The MSP methods and electron microscopy would be excluded for such public use because of the length of time to get an answer and the somewhat special skill required in operating them. These limitations could be overcome by advancements in the technology and automated sampling.

PCR and immunological methods could be used in a similar manner. Antibody test kits are available and are released for home use. However, a positive result using a home kit comes with a recommendation that the test be confirmed by another test or methods elsewhere. Such methods are similar to a pregnancy test kit, and two tests that are positive does not mean you will have twins. A professional would need to follow-up.

8.6.2 Fixed sites

One way to address this public use is the utilization of conveniently located testing centers. Such centers already exist in most locations and are used by local hospitals, physicians, and others to collect and analyze such samples. These centers could easily be used for virus detection. Which technology to use routinely can then be determined. Again, it would appear that the IVDS instrument could be used for testing for a virus. Another technology could be used to confirm the virus and/or make an identification: immunological, PCR, MSP, electron microscopy. Remembering that a positive detection by IVDS may not necessarily be confirmed by PCR and immunology, because of the limit in viruses primed to detect, and the error rate of over 60%, other technologies may be needed such as MSP or the electron microscope. Nevertheless, routine viruses can be detected and confirmed. IVDS can sort out the negatives and PCR and immunological methods identity. This combination is important simply because of the preponderance of negatives and the impracticable need to keep and store a wide range of kits. Where only a few viruses are indicated, then a multiplex system or several test kits could be used for verification. It is not recommended to just use the PCR and immunological kits as a primary means for testing for viruses, because of their high error rate and the virus + one rule, where it can be likely a false negative (where the method is blind) is reported and the person is actually positive for a virus. This is particularly important when considering a new virus.

8.6.3 Protecting small high value groups

Consider a person or group that is important and that cannot be exposed to a virus. This group may be government officials, city officials, governing bodies, or police and emergency responders. It can be a common virus or an exotic unknown virus. How do you make sure? One approach is to test everyone who comes into contact with this person or group. We know the failure of using only PCR and immunological methods by the fact that heads of state have contracted COVID-19. This leads to the question of other viruses or indeed a new virus beyond our experience. There is no excuse for "almost pregnant". Test everyone, at the door, with IVDS. It can be used continuously, 24/7, with a minimum of supplies. Personnel can be tested frequently and there is no resupply or shelf-life problem. Frequency of testing can be determined as needed from high to low without having to be concerned about having enough test kits or the right test kits. Since a sample test takes about 5 minutes for saliva it is not invasive. People with a positive test are not admitted and seek assistance. Their test is then subject to confirmation by other means.

8.6.4 Protecting small groups

Numerous small groups such as on airlines, trains, and buses as well as those attending conferences, international groups, meetings, and similar groupings of people coming together from some distance or from different environments indicate attention. It is not enough to ask a few questions and wear a surgical mask to protect the participants. Use a mobile testing capability. Instruments indicated are IVDS to detect the present or absence of a virus followed by PCR and immunological methods to confirm the identity of the virus if possible. Nevertheless, people with a positive detection are not allowed into the meeting. Those in the meeting who have tested negative will have a high confidence of not being exposed to a virus.

People traveling are frequently tested for COVID-19 before they travel. Given the high error rate of these tests many positives get through, and may negatives end up not traveling. This is highly unacceptable; it is better to not test at all. A mobile testing platform is indicated. As with any group, use the IVDS to test for the presence or absence of a virus and PCR and immunological methods to confirm for a list of known viruses. Several mobile units can come together to test larger groups. In this manner test all travelers on all sorts of transportation.

The mobile units have been placed in a vehicle. The testing process can be conducted for schools (hundreds of people), departments, and businesses.

8.6.5 Protecting large groups

Bringing together several mobile testing stations works for a few hundred or maybe a few thousand people. Where people gather for a sporting event, test at the entrance. Mobile units could be available and test everyone at least once. The number to be tested and the time allowed would determine the number of mobile units. Testing hundreds of thousands or millions in other conditions poses different challenges. Statistical sampling and other processes are indicated.

References

AOAC International. 2004. "Initiative yields effective methods for anthrax detection; RAMP and MIDI, Inc., methods approved." *Inside Laboratory Management* 10(3).

Arnold, R. J. and J. P. Reilly. 1999. "Observation of Escherichia coli ribosomal proteins and their posttranslational modifications by mass spectrometry." *Analytical Biochemistry* 269(1):105–12.

Baillie, L. W., M. N. Jones, P. C. Turnbull, and R. J. Manchee. 1995. "Evaluation of the BIOLOG® system for the identification of Bacillus anthracis." *Letters in Applied Microbiology* 20(4):209–11.

Bakhtiar, R., and Z. Guan. 2005. "Electron capture dissociation mass spectrometry in characterization of post-translational modifications." *Biochemical and Biophysical Research Communications* 334(1):1–8.

Barinaga, C. J., D. W. Koppenaal, S. A. McLuckey. 1994. "Ion-trap mass spectrometry with an inductively coupled plasma source." *Rapid Communications in Mass Spectrometry* 8(1):71–76.

Baron, C., N. Domke, R. M. Beinhofe, and M. S. Hapfel. 2001. "Elevated temperature differentially affects virulence, VirB protein accumulation, and T-pilus formation in different Agrobacterium tumefaciens and Agrobacterium vitis strains." *Journal of Bacteriology* 183(23):6852–6861.

Bell, C. A., J. R. Uhl, T. L. Hadfield, et al. 2002. "Detection of bacillus anthracis DNA by lightcycler PCR." *Journal of Clinical Microbiology* 40(8):2897–2902.

Bertrand, M. J., P. Martin, and O. Peraldi. 2000. *A New Concept in Benchtop Mass Spectrometer, MAB-ToF.* New Orleans: PittCon.

Bolbach, G. 2005. "Matrix-assisted laser desorption/ionization analysis of non-covalent complexes: Fundamentals and applications." *Current Pharmaceutical Design* 11(20):2535–57.

Bothner, B., A. Schneemann, D. Marshall, V. Reddy, et al. 1999. "Crystallographically identical virus capsids display different properties in solution." *Nature Structural & Molecular Biology* 6:114–116.

Bouquet, H. 1763. *Bouquet Letters, MSS 21634:295 and 231.* London: British Library.

Bromenshenk, J.J, C. B. Henderson, C. H.Wick, et al. 2010. "Iridovirus and microsporidian linked to honey bee colony decline." *PLoS ONE* 5(10): doi:10.1371/journal.pone.0013181

Bruins, A. P. 1991. "Liquid chromatography-mass spectrometry with ionspray and electrospray interfaces in pharmaceutical and biomedical research." *Journal of Chromatography* 554(1–2):39–46.

Cargile, B. J., S. A. Mcluckey, and J. L. Stephenson. 2001. "Identification of bacteriophage MS2 coatprotein from E. coli lysates via ion trap collisional activation of intact protein ions." *Analytical Chemistry* 73:1277–1285.

Chait, B. T. and S. B. H. Kent. 1992. "Weighing naked proteins: Practical, high-accuracy mass measurement of peptides and proteins." *Science* 257(5078):1885–94.

Chen, R., X. Cheng, D. W. Mitchell, et al. 1995. "Trapping, detection, and mass determination of Coliphage T4 DNA Ions by electrospray ionization Fourier transform ion cyclotron resonance mass spectrometry." *Analytical Chemistry* 67(7):1159–63.

Chen, S. 1997. "Tandem mass spectrometric approach for determining structure of molecular species of amino phospholipids." *Lipids* 32(1):85–100.

Chen, W., K. E. Laidig, Y. Park, et al. 2001. "Searching the Porphyromonas gingivalis genome with peptide fragmentation mass spectra." *Analyst* 126:52–57.

Chenna, A., and C. R. Iden. 1993. "Characterization of 2'-deoxycytidine and 2'-deoxyuridine adducts formed in reactions with acrolein and 2-bromoacrolein." *Chemical Research in Toxicology* 6(3): 261–68.

Cohen, S. L., and B. T. Chait. 1997. "Mass spectrometry of whole proteins eluted from sodium dodecyl sulfate-polyacrylamide gel electrophoresis gels." *Analytical Biochemistry* 247: 257–67.

Cotter, R. J. 1992. "Time-of-flight mass spectrometry for the structural analysis of biological molecules." *Analytical Chemistry* 64(21):A1027–A1039.

Dai,Y., L. Li, D. C. Roser, and S. R. Long. 1999. "Detection and identification of low-mass peptides and proteins from solvent suspensions of Escherichia coli by high performance liquid chromatography fractionation and matrix-assisted laser desorption/ionization mass spectrometry." *Rapid Communications in Mass Spectrometry* 13:73–78.

Dawson, P. H. 1976. *Quadrupole Mass Spectrometry and its Application.* Amsterdam: Elsevier.

Doroshenko, V. M., and R. J. Cotter. 1994. "Linear mass calibration in the quadrupole ion-trap mass spectrometer." *Rapid Communications in Mass Spectrometry* 8(9):766–71.

Doroshenko, V. M., and R. J. Cotter. 1996. "Advanced stored waveform inverse Fourier transform technique for a matrix-assisted laser desorption/ionization quadrupole ion trap mass spectrometer." *Rapid Communications in Mass Spectrometry* 10(1):65–73.

Dumas, M.-E., L. Debrauwer, L. Beyet, et al. 2002. "Analyzing the physiological signature of anabolic steroids in cattle urine using pyrolysis/metastable atom bombardment mass spectrometry and pattern recognition." *Analytical Chemistry* 74(20):5393–5404.

Dworzanski, J. P., A. P. Snyder, R. Chen, et al. 2004. "Correlation of mass spectrometry identified bacterial biomarkers from a fielded pyrolysis-gas chromatography-ion mobility spectrometry biodetector with the microbiological gram stain classification scheme." *Analytical Chemistry* 76(21):2355–66.

Dworzanski, J. P., S. V. Deshpande, R. Chen, et al. 2006. "Mass spectrometry-based proteomics combined with bioinformatics tools for bacterial classification." *Journal of Proteome Research* 5(1):76–87.

Ecelberger, S. A., T. J. Cornish, B. F. Collins, et al. 2004. "Suitcase TOF: A man-portable time-of-flight mass spectrometer." *Johns Hopkins APL Technical Digest* 25(1):14–19.

Fader, R. C., L. K. Duffy, C. P. Davis, and A. Kurosky. 1982. "Purification and chemical characterization of type I pili isolated from Klebsiella pneumonia." *Journal of Biological Chemistry* 25:257(6):3301–3305.

Fernandez, L. E., H. R. Sorensen, C. Jorgensen, et al. 2007. "Characterization of oligosaccharides from industrial fermentation residues by matrix-assisted laser desorption/ionization, electrospray mass spectrometry, and gas chromatography mass spectrometry." *Molecular Biotechnology* 35(2):149–60.

Fleischmann, R., M. Adams, O. White, et al. 1995. "Whole-genome random sequencing and assembly of Haemophilus Influenza Rd." *Science* 269:496–512.

Fountain, S. T., H. Lee, and D. M. Lubman. 1994. "Ion fragmentation activated by matrix-assisted laser desorption/ionization in an ion-trap/reflectron time-of-flight device." *Rapid Communications in Mass Spectrometry* 8(5):487–94.

Fox, A., and K. Fox. 1991. "Rapid elimination of a synthetic adjuvant peptide from the circulation after systemic administration and absence of detectable natural muramyl peptides in normal serum at current analytical limits." *Infectection and Immunity* 59(3):1202–14.

Fox, A., G. E. Black, K. Fox, and R. Rostovtseva. 1993. "Determination of carbohydrate profiles of Bacillus anthracis and Bacillus cereus including identification of O-methyl methylpentoses by using gas chromatography-mass spectrometry." *Journal of Clinical Microbiology* 31(4):887–94.

Gabelica, V., C. Vreuls, P. Filee, et al. 2002. "Advantages and drawbacks of nanospray for studying noncovalent protein-DNA complexes by mass spectrometry." *Rapid Communications in Mass Spectrometry* 16(18):1723–28.

Gale, D. C., J. E. Bruce, G. A. Anderson, et al. 1993. "Bio-affinity characterization mass spectrometry." *Rapid Communications in Mass Spectrometry* 7:1017–21.

Guilhaus, M. 1995. "Principles and instrumentation for TOF-MS." *Journal of Mass Spectrometry* 30(11):1519–32.

Hager, J. W., and J. C. Le Blanc. 2003. "High-performance liquid chromatography-tandem mass spectrometry with a new quadrupole/linear ion trap instrument." *Journal of Chromatography A* 1020(1):3–9.

Harris, S. 1992. *Chemical and Biological Warfare, Ann.* New York: New York Academy of Sciences. 666:21–48.

Hillenkamp, F., M. Karas, R. C. Beavis, and B. T. Chait. 1991. "Matrix-assisted laser desorption/ionization mass spectrometry of biopolymers." *Analytical Chemistry* 63(24):A1193–A1202.

Hopfgartner, G., C. Husser, and M. Zell. 2003. "Rapid screening and characterization of drug metabolites using a new quadrupole-linear ion trap mass spectrometer." *Journal of Mass Spectrometry* 38(2):138–50.

Hopfgartner, G., E. Varesio, V. Tschäppät, et al. 2004. "Triple quadrupole linear ion trap mass spectrometer for the analysis of small molecules and macromolecules." *Journal of Mass Spectrometry* 39(8):845–55.

Hsu, J., S. J. Chang, and A. H. Franz. 2006. "MALDI-TOF and ESI-MS analysis of oligosaccharides labeled with a new multifunctional oligosaccharide tag." *Journal of the American Society for Mass Spectrometry* 17(2):194–204.

Hu, Q., R. Noll, H. Li, et al. 2005. "The Orbitrap: A new mass spectrometer." *Journal of Mass Spectrometry* 40(4):430–43.

Hua, Y., W. Lu, M. S. Henry, et al. 1993. "Online high-performance liquid chromatography-electrospray ionization mass spectrometry for the determination of brevetoxins in 'Red Tide' algae." *Analytical Chemistry* 67(11):1815–1823.

Hunt, D. F., and F. W. Crow. 1978. "Electron capture negative ion CI mass spectrometry." *Analytical Chemistry* 50(13):1781–84.

Hutchens, T. W., and T. T. Yip. 1993. "New desorption strategies for the mass spectrometric analysis of macromolecules." *Rapid Communications in Mass Spectrometry* 7(7):576–80.

Ibekwe, A. M., and C. M. Grieve. 2003. "Detection and quantification of Escherichia coli O157:H7 in environmental samples by real-time PCR." *Journal of Applied Microbiology* 94(3):421–31.

Ivnitski, D., D. J. O'Neil, A. Gattuso, et al. 2003. "Nucleic acid approaches for detection and identification of biological warfare and infectious disease agents." *BioTechniques* 35:862–69.

Jensen, P. K., L. Pasa-Tolic, G. A. Anderson, et al. 1999. "Probing proteomes using capillary Isoelectric focusing-electrospray ionization fourier transform ion cyclotron resonance mass spectrometry." *Analytical Chemistry* 71:2076–2084.

Jonscher, K. R., and J. R. Yates, III. 1996. "Mixture analysis using a quadrupole mass filter/quadrupole ion trap mass spectrometer." *Analytical Chemistry* 68(4):659–67.

Karas, M., and F. Hillenkamp. 1988. "Laser desorption ionization of proteins with molecular masses exceeding 10,000 daltons." *Analytical Chemistry* 60(20):259–80.

Karataev, V. I., B. A. Mamyrin, and D. V. Shmikk. 1972. "New method for focusing ion bunches in time-of-flight mass spectrometers." *Soviet Physics - Technical Physics* 16:1177–79.

Karty, J. A., S. Lato, and J. P. Reilly. 1998. "Detection of the bacteriological sex factor in E. coli by matrix-assisted laser desorption/ionization time-of-flight mass spectrometry." *Rapid Communications in Mass Spectrometry* 12:625–629.

King, R., and C. Fernandez-Metzler. 2006. "The use of QTrap technology in drug metabolism." *Current Drug Metabolism* 7(5):541–45.

Klietmann, W. F., and K. L. Ruoff. 2001. "Bioterrorism: Implications for the clinical microbiologist." *Clinical Microbiology Reviews* 14(2):364–81.

Krishnamurthy, T., U. Rajamani, P. L. Ross, R. Jabbour, et al. 2000. "Mass spectral investigations on microorganisms." *Journal of Toxicology - Toxin Reviews* 19:95–117.

Kuzmanovic, D. A., I. Elashvili, C. H. Wick, et al. 2003. "Bacteriophage MS2: Molecular weight and spatial distribution of the protein and RNA components by small-angle neutron scattering and virus counting." *Structure* 11:1339–1348.

Laiko, V. V., S. C. Moyer, and R. J. Cotter. 2000. "Atmospheric pressure MALDI/ion trap mass spectrometry." *Analytical Chemistry* 72(21):5239–43.

Lei, Q. P., X. Cui, D. M. J. Kurtz, et al. 1998. "Electrospray mass spectrometry studies of non-heme iron-containing proteins." *Analytical Chemistry* 70:1838–1846.

Li, J., Z. Zhang, J. Rosenzweig, et al. 2002. "Proteomics and bioinformatics approaches for identification of serum biomarkers to detect breast cancer." *Clinical Chemistry* 48(8):1296–1304.

Li, Y., F. Wenzel, W. Holzgreve, and S. Hahn. 2006. "Genotyping fetal paternally inherited SNPs by MALDI-TOF MS using cell-free fetal DNA in maternal plasma: Influence of size fractionation." *Electrophoresis* 27(19):3889–96.

Linde, Hans-Jörg, H. Neubauer, H. Meyer, et al. 1999. "Identification of Yersinia species by the Vitek GNI card." *Journal of Clinical Microbiology* 37(1):211–14.

Liu, J., K. W. Ro, M. Busman, and D. R. Knapp. 2004. "Electrospray ionization with a pointed carbon fiber emitter." *Analytical Chemistry* 76(13):3599–3606.

Liu, Tsang-Yu, Lung-Lin Shiu, Tien-Yau Luh, and Guor-Rong Her. 1995. "Direct analysis of C60 and related compounds with electrospray mass spectrometry." *Rapid Communications in Mass Spectrometry* 9(1): 93–96.

Macek, B., L. F. Waanders, J. V. Olsen, and M. Mann. 2006. "Top-down protein sequencing and MS3 on a hybrid linear quadrupole ion trap-orbitrap mass spectrometer." *Molecular & Cellular Proteomics* 5(5): 949–58.

Makarov, A., E. Denisov, A. Kholomeev, et al. 2006b. "Performance evaluation of a hybrid linear ion trap/orbitrap mass spectrometer." *Analytical Chemistry* 78(7):2113–20.

Makarov, A., E. Denisov, O. Lange, and S. Horning. 2006a. "Dynamic range of mass accuracy in LTQ Orbitrap hybrid mass spectrometer." *Journal of the American Society for Mass Spectrometry* 17(7):977–82.

Mamyrin, B. A., V. I. Karataev, D. V. Shmikk, and V. A. Zagulin. 1973. "Reflectron for TOF-MS." *Soviet Physics Journal of Experimental and Theoretical Physics* 37:45–48.

Mamyrin, B. A. 2001. "Time-of-flight mass spectrometry (concepts, achievements and prospects)." *International Journal of Mass Spectrometry* 206(3):251–66.

Mann, M., and Wilm, M. S. 1994. "Error-tolerant identification of peptides in sequence databases by peptide sequence tags." *Analytical Chemistry* 66(24):4390–99.

Martin, S. E., J. Shabanowitz, D. F. Hunt, and J. Marto. 2000. "Subfemtomole MS and MS/MS peptide sequence analysis using nano-HPLC micro-ESI Fourier transform ion cyclotron resonance mass spectrometry." *Analytical Chemistry* 72(18):4266–74.

McCluckey, S. A., G. Vaidyanathan, and S. Habibi-Goudarzi. 1995. "Charged vs. neutral nucleobase loss from multiply charged oligonucleotide anions." *Journal of Mass Spectrometry* 30(9):1222–29.

Morris, H. R., T. Paxton, A. Dell, et al. 1996. "High sensitivity collisionally-activated decomposition tandem mass spectrometry on a novel quadrupole/orthogonal-acceleration time-of-flight mass spectrometer." *Rapid Communications in Mass Spectrometry* 10(8):889–96.

Morse, S. S., ed. 1993. *Emerging Viruses*. New York: Oxford University Press.

Mouradian, S., et al. 1997. "DNA analysis using an electrospray scanning mobility particle sizer." *Analytical Chemistry* 69:919–925.

Murray, P. R., E. J. Baron, J. H. Jorgensen, M. A. Pfaller, and R. H. Yolken, eds. 2003. *Manual of Clinical Microbiology*. 8th ed. Washington, DC: ASM Press.

Odumeru, J. A., M. Steele, L. Fruhner, et al. 1999. "Evaluation of accuracy and repeatability of identification of food-borne pathogens by automated bacterial identification systems." *Journal of Clinical Microbiology* 37(4):944–49.

Owens, D. R., B. Bothner, Q. Phung, K. Harris, and G. Siuzdak. 1998. "Aspects of oligonucleotide and peptide sequencing with MALDI and electrospray mass spectrometry." *Bioorganic & Medicinal Chemistry* 6(9):1547–54.

Qian, M. G., and D. M. Lubman. 1995. "Analysis of tryptic digests using microbore HPLC with an ion trap storage/reflectron time-of-flight detector." *Analytical Chemistry* 67(17): 2870–77.

Schneider, B. B., V. I. Baranov, H. Javaheri, and T. R. Covey. 2003. "Particle discriminator interface for nanoflow ESI-MS." *Journal of the American Society for Mass Spectrometry* 14(11):1236–46.

Shukla, A. K., and J. H. Futrell. 2000. "Tandem mass spectrometry: Dissociation of ions by collisional activation." *Journal of Mass Spectrometry* 35(9):1069–90.

Skoog, D. A., F. J. Holler, and T. A. Nieman, eds. 1992. *Principles of Instrumental Analysis*. 5th ed. Orlando: Saunders College Publishing, 184.

Sleno, L., and D. A. Volmer. 2004. "Ion activation methods for tandem mass spectrometry." *Journal of Mass Spectrometry* 39(10):1091–1112.

Smith, P. B. W., A. P. Snyder, and C. S. Harden. 1995. "Characterization of bacterial phospholipids by electrospray ionization tandem mass spectrometry." *Analytical Chemistry* 67(11):1824–30.

Smith, R. D., X. Cheng, J. E. Bruce, et al. 1994. "Trapping, detection and reaction of very large single molecular ions by mass spectrometry." *Nature* 369:137–39.

Speir, J. P., G. S. Gorman, C. C. Pitsenberger, et al. 1993. "Remeasurement of ions using quadrupolar excitation Fourier transform ion cyclotron resonance spectrometry." *Analytical Chemistry* 65(13):1746–52.

St. Geme, J. W., J. S. Pinkne, G. P. Krasan, et al. 1996. "Haemophilus infiuenzae pili are composite structures assembled via the HifB chaperone." *Proceedings of the National Academy of Sciences* 93:11913–11918.

Stimson, E., M. Virji, S. Barker, M. Panico, et al. 1996. "Discovery of a novel protein modification: Alpha-glycerophosphate is substituent of meningococcal pilin." *Biochemical Journal* 315:29–33.

Stockholm International Peace Research Institute. 1971. *The Problem of Chemical and Biological Warfare*. Volume 1. New York: Humanities Press.

Sun, X., and B. Guo. 2006. "Genotyping single-nucleotide polymorphisms by matrix-assisted laser desorption/ionization time-of-flight-based mini-sequencing." *Methods in Molecular Medicine* 128:225–30.

Syka, J. E. P., J. J. Coon, M. J. Schroeder, et al. 2004. "Peptide and protein sequence analysis by electron transfer dissociation mass spectrometry." *Proceedings of the National Academic of Sciences of the United States of America* 101(26):9528–33.

Syka, John E. P., J. A. Marto, D. L. Bai, et al. 2004. "Novel linear quadrupole ion trap/FT mass spectrometer: Performance characterization and use in the comparative analysis of histone H3 post-translational modifications." *Journal of Proteome Research* 3(3):621–26.

Tang, N., P. Tornatore, and S. R. Weinberger. 2003. "Current developments in SELDI affinity technology." *Mass Spectrometry Reviews* 23(1):34–44.

Temin, H. M. 1993. "The high rate of retrovirus variation results in rapid evolution." In *Emerging Viruses edited by S.S. Morse, 219–225*. New York: Oxford University Press.

Thomas, J. J., B. Bothner, J. Traina, et al. 2004. "Electrospray ion mobility spectrometry of intact viruses. *Spectroscopy* 18:31–36.

Thomas, J. J., B. Falk, C. Fenselau, et al. 1998. "Viral characterization by direct analysis of capsid proteins." *Analytical Chemistry* 70:3863–3867.

Tito, M. A., T. Kasper, V. Karin, et al. 2000. "Electrospray time-of-flight mass spectrometry of the intact MS2 virus capsid." *Journal of the American Chemical Society* 122:3550–3551.

Todd, John F. J. 2005. "Ion trap mass spectrometer - past, present, and future (?)" *Mass Spectrometry Reviews* 10(1):3–52.

Tong, W., A. Link, J. K. Eng, and J. R. Yates, J. R. 1999. "Identification of proteins in complexes by solid-phase microextraction/multistep elution/capillary electrophoresis/tandem mass spectrometry." *Analytical Chemistry* 71:2270–2278.

Trainor, J. R., and P. J. Derrick. 1992. *Mass Spectrometry in the Biological Sciences: A Tutorial.* Dordrecht, Netherlands: Kluwer Academic Publishers:3–27.

Ushinsky, S. C., H. Bussey, A. A. Ahmed, et al. 1997. "Histone HI in Saccharomyces cerevisiae." *Yeast* 13:151–161.

Valaskovic, G. A., L. Utley, M. S. Lee, and J. T. Wu. 2006. "Ultra-low flow nanospray for the normalization of conventional liquid chromatography/mass spectrometry through equimolar response: Standard-free quantitative estimation of metabolite levels in drug discovery." *Rapid Communications in Mass Spectrometry* 20(7):1087–96.

Valegard, K., L. Liljas, K. Fridborg, and T. Unge. 1990. "The three-dimensional structure of the bacterial virus MS2." *Nature* 345:36–44.

Vanrobaeys, F., R. Van Costerk, G. Dhondt, et al. 2005. "Profiling of myelin proteins by 2D-gel electrophoresis and multidimensional liquid chromatography coupled to MALDI TOF-TOF mass spectrometry." *Journal of Proteome Research* 4(6):2283–93.

Vestal, M. L., P. Juhasz, and S. A. Martin. 1995. "Delayed extraction matrix-assisted time-of-flight mass spectrometry." *Rapid Communications in Mass Spectrometry* 9(11):1044–50.

Vorm, O., and P. Roepstorff. 1994. "Peptide sequence information derived by partial acid hydrolysis and matrix-assisted laser desorption/ionization mass spectrometry." *Biological Mass Spectrometry* 23(12):734–40.

Wang, J., R. S. Houk, D. Dreessen, and D. R. Wiederin. 1999. "Speciation of trace elements in proteins in human and bovine serum by size exclusion chromatography and inductively coupled plasma-mass spectrometry with a magnetic sector mass spectrometer." *Journal of Biological Inorganic Chemistry* 4(5):546–53.

Webster, R. G. 1993. "Influenza." In *Emerging Viruses*, edited by S. S. Morse, 37–45. New York: Oxford University Press.

Westin, L., C. Miller, D. Vollmer, et al. 2001. "Antimicrobial resistance and bacterial identification utilizing a microelectronic chip array." *Journal of Clinical Microbiology* 39(3):1097–1104.

Wick, C.H, 2015, *Integrated Virus Detection*, New York: CRC Press.

Wick, C. H, I. Elashvili, P. E. McCubbin, and A. Birenzvige. 2005. *Determination of MS2 Bacteriophage Stability at High Temperatures using the IVDS. ECBC-TR-453.* Aberdeen Proving Ground: U.S. Army Edgewood Chemical Biological Center.

Wick, C. H, and D. A. Wick. 2021. *Microbial Diversity in Honneybees*. Boca Raton. FL: CRC Press.

Wick, C. H. and D. M. Anderson. 2000. "System and method for detection identification and monitoring of submicron sized particles." U.S. Patent 6,051,189.

Wick, C. H. and P. E. McCubbin. 1999. "Purification of MS2 bacteriophage from complex growth media and resulting analysis by the integrated virus detection system (IVDS)." *Toxicology Methods* 9:253–263.

Wick, C. H., and P. E. McCubbin. 2005. *Stability of IVDS Electrospray Module during Analysis of MS2 Bacteriophage. ECBC-TR-462.* Aberdeen Proving Ground: U.S. Army Edgewood Research, Development and Engineering Center.

Wick, C. H., and P. E. McCubbin. 1999. "Characterization of purified MS2 bacteriophage by the physical counting methodology used in the integrated virus detection system (IVDS)." *Toxicology Methods* 9:245–252.

Wick, C. H., and P. E. McCubbin. 1999. "Passage of MS2 bacteriophage through various molecular weight filters." *Toxicology Methods* 9:265–273.

Wick, C. H., D. M. Anderson, and P. E. McCubbin. 1999. *Characterization of the Integrated Virus Detection System (IVDS) using MS2 Bacteriophage. ECBC-TR-018 (AD-A364117).* Aberdeen Proving Ground: U.S. Army Edgewood Chemical Biological Center.

Wick, C. H. 2002a. "Method and apparatus for counting submicron sized particles." U.S. Patent 6,485,686 Bl.

Wick, C. H. 2002b. "Method and system for detecting and recording submicron sized particles." U.S. Patent 6,491,872.

Wick, C. H. 2007. "Method and system for detecting and recording submicron sized particles." U.S. Patent 7,250,138 B2.

Wick, C. H. 2010. "Detecting bacteria by direct counting of structural protein units or pili by IVDS and mass spectrometry." U.S. Patent 7,850,908 Bl.

Wick, C. H. 2011. "Detecting bacteria by direct counting of structural protein units or pili and mass spectrometry." U.S. Patent 8,021,884.

Wick, C. H. 2012a. "Concentrator device and method of concentrating a liquid sample." U.S. Patent 8,146,446 Bl.

Wick, C. H. 2012b. "Virus and particulate separation from solution." U.S. Patent 8,309,029 Bl.

Wick, C. H. 2013a. "Method and system for sampling and separating submicron-sized particles based on density and or size to detect the presence of a particular agent." U.S. Patent 8,524,482 B1.

Wick, C. H. 2013b. "Virus and particulate separation from solution." U.S. Patent 8,524,155 B1.

Wick, C. H. and P. E. McCubbin. 2010. *Capillary Diameter Variation for the Integrated Virus Detection System (IVDS). ECBC-TR-811.* Aberdeen Proving Ground: U.S. Army Edgewood Research, Development and Engineering Center.

Wick, C. H. and P. E. McCubbin. 2010. *Malvern Nano ZS Particle Size Comparison with the Integrated Virus Detection System (IVDS). ECBC-TR-749.* Aberdeen Proving Ground: U.S. Army Edgewood Research, Development and Engineering Center.

Wick, C. H. and P. E. McCubbin. 2010. *Recovery of Virus Samples from Various Surfaces with the Integrated Virus Detection System. ECBC-TR-816.* Aberdeen Proving Ground: U.S. Army Edgewood Research, Development and Engineering Center.

Wick, C. H., ed. 2014. *Identifying Microbes by Mass Spectrometry Proteomics*. New York: CRC Press.

Wick, C. H., H. Carlon, H. Yeh, and D. Anderson. 1998. *Quasi-Real-time Monitor for Airborne Viruses. ERDEC-TR–45*. Aberdeen Proving Ground: U.S. Army Edgewood Chemical Biological Center.

Wick, C. H., H. R. Carlon, R. L. Edmonds, and J. Blew. 1997. *Rapid Identification of Airborne Biological Particles by Flow Cytometry, Gas Chromatography, and Genetic Probes. ERDEC-TR-443*. Aberdeen Proving Ground: U.S. Army Edgewood Chemical Biological Center.

Wick, C. H., H. R. Yeh, H. R. Carlon, and D. Anderson. 1997. *Virus Detection: Limits and Strategies. ERDEC-TR-453*. Aberdeen Proving Ground: U.S. Army Edgewood Research, Development and Engineering Center.

Wick, C. H., I. Elashvili, P.E. McCubbin, and A. Birenzvige. 2005. *Determination of MS2 Bacteriophage Stability at High pH using the IVDS. ECBC-TR-472*. Aberdeen Proving Ground: U.S. Army Edgewood Chemical Biological Center.

Wick, C. H., I. Elashvili, R. Jabbour, et al. 2006. "Mass spectrometry and integrated virus detection system characterization of MS2 bacteriophage." *Toxicology Mechanisms and Methods* 16:485–493.

Wick, C. H., M. D. Dunkel, R. Crumley, et al. 1998. *Pulsed Light Device (PLD) for Deactivation of Biological Aerosols. ERDEC-TR-456*. Aberdeen Proving Ground: U.S. Army Edgewood Chemical Biological Center.

Wick, C. H., M. M. Wade, T. D. Biggs, et al. 2012. *Effects of Repeated Exposure to Filtered and Unfiltered Broadband Light Radiation on Escherichia coli Growth and Propagation. ECBC-TR-987*. Aberdeen Proving Ground: U.S. Army Edgewood Research, Development and Engineering Center.

Wick, C. H., P. E. McCubbin, and A. Birenzvige. 2005. *Detection and Identification of Viruses Using the Integrated Virus Detection System (IVDS). ECBC-TR-463 (AD-A454 377)*. Aberdeen Proving Ground: U.S. Army Edgewood Research, Development and Engineering Center.

Wick, C. H., P. E. McCubbin, and A. Birenzvige. 2006. *Determination of MS2 Bacteriophage stability At High pH Using the IVDS. ECBC-TR-472*. Aberdeen Proving Ground: U.S. Army Edgewood Research, Development and Engineering Center.

Wick, C. H., P. E. McCubbin, and A. Birenzvige. 2006. *Determination of MS2 Bacteriophage stability at Low pH using the IVDS. ECBC-TR-473*. Aberdeen Proving Ground: U.S. Army Edgewood Research, Development and Engineering Center.

Wick, C. H., R. L. Edmonds, and J. Blew. 1995. *Rapid Detection and Identification of Background Levels of Airborne Biological Particles. ERDEC-TR-155.* Aberdeen Proving Ground: U.S. Army Edgewood Chemical Biological Center.

Wick, C. H., S. Weugraitis, and P. E. McCubbin. 2010. *Integrated Virus Detection System Characterization of MS2 and TBSV after Pulsed Light Exposure. ERBC-TR-817.* Aberdeen Proving Ground: U.S. Army Edgewood Research, Development and Engineering Center.

Wickman, G., B. Johansson, J. Bahar-Gogani, et al. 1998. "Liquid ionization chambers for absorbed dose measurements in water at low dose rates and intermediate photon energies." *Medical Physics* 25(6):900–907.

Wilm, M. S., and M. Mann. 1996. "Analytical properties of the nanoelectrospray ion source." *Analytical Chemistry* 68(1):1–8.

Wong, P. S., N. Srinivasan, N. Kasthurikrishnan, et al. 1996. "On-line monitoring of the photolysis of benzyl acetate and 3,5-dimethoxybenzyl acetate by membrane introduction mass spectrometry." *Journal of Organic Chemistry* 61(19):6627–32.

Wright, Jr., G. L., L. H. Cazares, S. M. Leung, et al. 1999. "Proteinchip® surface enhanced laser desorption/ionization (SELDI) mass spectrometry: A novel protein biochip technology for detection of prostate cancer biomarkers in complex protein mixtures." *Prostate Cancer Prostatic Diseases* 2:264–76.

Wu, J., and S. A. McLuckey. 2003. "Ion/ion reactions of multiply charged nucleic acid anions: Electron transfer, proton transfer, and ion attachment." *International Journal of Mass Spectrometry* 228(2–3):577–97.

Wuhrer, M., and A. M. Deelder. 2006. "Matrix-assisted laser desorption/ionization in-source decay combined with tandem time-of-flight mass spectrometry of permethylated oligosaccharides: Targeted characterization of specific parts of the glycan structure." *Rapid Communications in Mass Spectrometry* 20(6):943–51.

Yanes, O., J. Villanueva, E. Querol, and F. X. Aviles. 2007. "Detection of non-covalent protein interactions by 'intensity fading' MALDI-TOF mass spectrometry: Applications to proteases and protease inhibitors." *Nature Protocols* 2:119–30.

Yates, J. R. 1998. "Mass spectrometry and the age of the proteome." *Journal of Mass Spectrometry* 33(1):1–19.

Yost, R. A., and C. G. Enke. 1979. "TQMS for mixture analysis." *Analytical Chemistry* 51:1251A–64A.

Yost, R. A., and C. G. Enke. 1983. "TQMS." *In Tandem Mass Spectrometry*, edited by F. W. McLafferty, 175–96. New York: Wiley.

Zambonin, C. G., C. D. Calvano, L. D'Accolti, and F. Palmisano. 2006. "Laser desorption/ionization time-of-flight mass spectrometry of squalene in oil samples." https://pubmed.ncbi.nlm.nih.gov/16345138

Zerega, Y., J. A. G. Brincourt, and R. Catella. 1994. "A new operating mode of a quadrupole ion trap in mass spectrometry: Part 1. Signal visibility." *International Journal of Mass Spectrometry and Ion Processes* 132(1–2):57–72.

Zhang, J., K. Schubothe, B. Li, et al. 2005. "Infrared multiphoton dissociation of O-linked mucin-type oligosaccharides." *Analytical Chemistry* 77(1):208–14.

Zubarev, R. A., N. L. Kelleher, and F. W. McLafferty. 1998. "Electron capture dissociation of multiply charged protein cations: A nonergodic process." *Journal of the American Chemical Society* 120(13):3265–66.

Zubarev, R. A. 2004. "Electron-capture dissociation tandem mass spectrometry." *Current Opinion in Biotechnology* 15(1):12–16.

Index

A

Acquired immunodeficiency syndrome (AIDS), 12
Aedes aegypti mosquito, 13
African swine fever virus (ASFV), 85, 87–88
Agents of Biological Origin Identification (ABOid), 29, 83–85, 91–92
Airborne viruses, 65
Air scrubbing, 65
Alcelaphine herpesvirus 1 (AlHV-1), 88
Amplification, 47
Ancient sewage systems, 1
Annealing temperature (T_a), 49
Antibody methods, 55
  adding new viruses, 55
  detecting viruses, 54–55
Antibody test kits, 115
Antonine Plague, 5
AOAC Official Methods of Analysis, 29
Arboviruses, 10, 13
Arthropod-borne virus, 13
Asian flu, 12
Aujeszky's disease, 88
Automated systems, 28

B

Bacillus anthracis, 29
Bacteria, 24
  detection/classification, 24–25
Bacteriology, 24
Bacteriophage sequences, 93
Beta instrument, 58
BioLog, 24
Biological warfare agents (BWAs), 29
Biological warfare (BW) environment, 64, 65
Biothreat agents, 27, 29, 32
Bluetongue disease, 15
Bovine malignant catarrhal fever (BMCF), 88

C

Cacao, 15
Camelpox virus (CMLV), 89
Campaign jaundice, 13
Capripoxviruses (CaPVs), 89
Cassava mosaic virus, 15, 16
Celsus, 9
Centrifugation, 61
Cercopithecine herpesvirus 5 (CeHV-5), 89
Chemical Biological (CB) MS system, 29
Chordopoxvirus (ChPV), 89
Citrus tristeza virus (CTV), 16
Clean samples, 58
Cocoliztli epidemics, 8
Complementary DNA (cDNA), 47, 51, 52
Contrast, 36
Coronavirus Data Group, 95
Coronavirus disease 2019, 18, 95
COVID-19, 18, 41, 94; *see also* Coronavirus disease 2019
  5 days and 3-month follow-up, 66–76
  antibody detection process, 55
  national unique peptide averages, 96
COVID-19 SARS-CoV-2 virus, 18
Creutzfeldt-Jakob disease, 64
Culex modestus, 17
Curl, 9
Cycle threshold (C_t), 50
Cytomegalovirus (CMV), 89

D

Dengue fever, 8, 40
Dengue virus, 13
Detection platform method, 21
Differential mobility analyzer (DMA) unit, 63, 72
Digital PCR, 50
Direct virus counting, 103

E

Ebola virus, 18, 64
Electron microscopes, 35, 108
Electron microscopy, 26, 35, 36, 66, 106
  scanning electron microscopy (SEM), 35, 38–43
    use, 38, 43
    working, 38, 40
  transmission electron microscopy (TEM), 35
    working, 36–37
  use of, 37
Electrospray (ES), 58
Electrospray ionization (ESI) method, 82–83
Electrospray mass spectrometer-mass spectrometer (ES-MS-MS), 79
End-point detection, 49
"The English sweat," 6
Enzyme-linked immunosorbent assay (ELISA) design, 28
Epizootics, 14, 17
Eupatorium lindleyanum, 5

F

Filoviruses, 18
Filterable virus, 9
Filters, 33
Fine-skinned *Bos taurus* cattle breeds, 90
Fistulina hepatica, 45
Food and Agriculture Organization (FAO), 15
Foot-and-mouth disease, 14
Fossil record, 1
Friesian System of Descriptive Taxonomy, 45
Fungal Tree of Life (AFTOL) Project, 45
Fungi, 21, 23, 45
  detection/classification, 23–24

G

Gas chromatography (GC), 29
Geminivirus, 16
GenBank: MN908947.3, 95
Genomic information, 26, 27
Germ theory of disease, 2
Global Polio Eradication Initiative, 12
Global Rinderpest Eradication Program, 15
Goatpox virus Pellor (GTPV), 89

Gram negative organism, 24
Gram positive organism, 24
GTPV strain Pellor (PL), 89

H

Hantavirus, 6, 7
Hepatitis, 13
Hepatitis A virus, 13
Hepatitis B virus, 13, 14
Herpes simplex virus (HSV), 42
Human Genome Project, 19
Human immunodeficiency virus (HIV), 12, 41
Human metapneumovirus, 10
Hydrophobia, 9

I

Image contrast mechanisms, 36
Immunoassays, 27, 28
Infantile paralysis, 11
Influenza epidemic, 5, 7
Influenza viruses, 5, 12
Instrument operation, 80
Integrated Virus Detection System (IVDS), 28, 30, 57, 58, 103, 106
- fielded, 77
- flow diagram for sample processing, 61
- improving sensitivity of, 66, 68
- invention, 58–60
- uses of, 62–66
- for virus detection, 60–62

International Committee on Taxonomy of Viruses (ICTV), 26
Ionization methods, 81
Ionization processes, 82
Irish Great Famine of 1845–1852, 9
Isothermal amplification, 28

J

Jian'an Plague, 5

K

Kivu Ebola epidemic 2018–2020, 17

L

La Crosse encephalitis, 13
Laws of Eshnunna, 9
Lumpy skin disease virus (LSDV), 89–90

M

Mad-cow disease, 64
Malignant catarrhal fever, 88
Man-made virus, 114
Marburg virus, 18
Mass spectral analysis, 81
Mass spectrometers, 29, 30, 79, 80
Mass spectrometry (MS) methods, 29, 32
Mass spectrometry proteomics (MSP) method, 25, 28, 30, 32, 46, 106
- detection and identification of viruses, 84–86
- national average for viruses, honeybees, 89, 90
- using ABOID, 91–92
  - African swine fever virus (ASFV), 87–88
  - *Alcelaphine herpesvirus 1* (AlHV-1), 88
  - camelpox virus (CMLV), 89
  - *Cercopithecine herpesvirus 5* (CeHV-5), 89
  - goatpox virus Pellor (GTPV), 89
  - lumpy skin disease virus (LSDV), 89–90
  - monkeypox virus Zaire-96-I-16 (MPV), 90
  - sheeppox virus (SPV), 90
  - vaccinia viruses (VACV), 90
  - variola virus (VARV), 90–91
- work for biological detection, 83–84

Mass-to-charge (m/z) ratio, 81
Measles virus, 5, 10
Melting temperature (T_m), 48, 49
Miasma, 2
Microbes, 3, 30, 31, 93
Microbial agents, 81
Microbial identification system (MIDI), 29
Microbiology, 21
MicroLog, 24–25
Microorganisms, 19, 31
MIDI BIOTER database, 29
MIDI Sherlock system, 29
Mobile units, 117
Molecular-based virus detectors, 46
Molecular ions, 81
Molecular methods, 27, 45
Monkeypox virus Zaire-96-I-16 (MPV), 90
MS2 bacteriophage, 58, 68
MSP-ABOID detection of viruses, 99
Multi-stage filtration, 61

N

n+1 rule, 54
National average for coronavirus
- COVID-19 detection discussion, 99
  - honeybee smoothie sample, 99
  - positive detection and identification of sample, 99
- verifying COVID-19 detection, 99

National Center for Biotechnology Information (NCBI), 84, 93
National Institutes of Health, 48
Negri bodies, 9
New viruses
- addition, 51
  - to antibody method of detection, 55
  - using MSP, 92–94
- identification, 38, 43
  - with direct counting, 58

Nigeria Lassa Fever epidemic, 17
Nipah virus, 17, 18
Nucleic acid–based methods, 27

P

Paramyxoviridae viruses, 53
Paramyxovirus, 18
Particle counting methods, 94
Phylogenetic mapping scheme, 21
Phylogenic tree, 30
Plague of Cyprian, 5
Plant viruses, 16
Poliomyelitis treatment, 11
Polio virus, 15
- infection, 5

Polymerase chain reaction (PCR), 4, 27, 32, 46–53
- high sensitivity of, 32
- instruments, 50–53
- *vs.* IVDS, 76–77

Potyvirus, 16
Primers, 47, 48, 52
Prokaryotic microorganisms, 24
Protein toxins, 31

Q

Quantum-mechanical phase shifts, 37

R

Rabies, 9
Rapid detection, 65, 106
RAW file, 83, 84, 92
Reagent-intensive approaches, 65
Reproducibility, 32
Reverse osmosis (RO) water, 66
Reverse transcriptase, 47, 51
Reverse transcription-polymerase chain reaction (RT-PCR), 47
Rhabdovirus, 9
Rice yellow mottle virus, 16
Rinderpest virus, 5, 6, 14, 15
RNA extraction, 51
Rodents, 90

S

SARS Data Group, 95
Scanning electron microscopy (SEM), 35, 38–43
  use, 38, 43
  working, 38, 40
Severe acute respiratory syndrome (SARS), 10, 18, 95
  coronavirus detection, 95
Severe acute respiratory syndrome coronavirus 2 (SARS-CoV-2), 18, 95
Sheeppox virus (SPV), 90
Sherlock Bioterrorism Library, 29
Small angle neutron scattering (SANS), 59
Smallpox eradication program, 10
Smallpox virus, 5, 7, 8, 10, 42
Snippet, 47, 48, 51
Spanish flu, 12
Specificity, 32
Spike protein, 66–67, 71–74

T

TaqMan probes, 27
Taq polymerase, 49, 52
Threshold line, 50
Tobacco mosaic virus (TMV), 33
Toxoptera citricidus, 16
Trade ships, 8
Transmission electron microscopy (TEM), 35
  working, 36–37
Trialeurodes vaporariorum, 15

U

Unknown microbes, 28, 53
US National Science Foundation, 45

V

Vaccinia viruses (VACV), 90
Variola Porcina, 87–88
Variola virus (VARV), 90
Virus Data Group, 93, 95
Virus detection methods, 102; *see also Individual entries*
  ability to detect multiple viruses, 106
  accurate detection, 104
  affordable, 104–106
  challenges, 108
    interference, 108
    sensitivity/trust in particular technology, 108–109
  clinical
    centralized testing center, 110
    clinic, 109
    hospital, 110
  environmental, 110
    agriculture, 110–112
    research, 112–113
    water, 112
  during pandemic, 113–114
  proven science, 103–104
  quick results (5–10 min), 106–107
  screens for unknown viruses, 106
  uses
    fixed sites, 115–116
    protecting large groups, 117
    protecting small groups, 116–117
    protecting small high value groups, 116
    public use, 115
Virus diseases, 9
Viruses, 25, 33, 58
  detection/classification, 25–29
Virus-free filters, 33
Virus Window, 57, 59, 63, 108
VITEK, 24

W

Waterborne sample, 31
West Nile virus, 13, 17–18
Wheat curl mite (*Aceria tulipae*), 16
Wheat streak mosaic virus, 16
World Animal Health Information Database (WAHID) Interface 2020, 91
World Health Assembly, 10
World Health Organization (WHO), 12, 91
World Organization for Animal Health, 15
Wuhan-Hu-1, 95

Y

Yellow fever, 13

Z

Zoonoses, 17
Zoonotic viral infections, 4, 17

9780367617981